Springer Tracts in Mechanical Engineering

Springer Tracts in Mechanical Engineering (STME) publishes the latest developments in Mechanical Engineering - quickly, informally and with high quality. The intent is to cover all the main branches of mechanical engineering, both theoretical and applied, including:

- Engineering Design
- Machinery and Machine Elements
- Mechanical structures and Stress Analysis
- Automotive Engineering
- Engine Technology
- Aerospace Technology and Astronautics
- Nanotechnology and Microengineering
- Control, Robotics, Mechatronics
- MEMS
- Theoretical and Applied Mechanics
- Dynamical Systems, Control
- Fluids mechanics
- Engineering Thermodynamics, Heat and Mass Transfer
- Manufacturing
- Precision engineering, Instrumentation, Measurement
- Materials Engineering
- Tribology and surface technology

Within the scopes of the series are monographs, professional books or graduate textbooks, edited volumes as well as outstanding PhD theses and books purposely devoted to support education in mechanical engineering at graduate and post-graduate levels.

Indexed by SCOPUS and Springerlink. The books of the series are submitted for indexing to Web of Science.

To submit a proposal or request further information, please contact: Dr. Leontina Di Cecco Leontina.dicecco@springer.com or Li Shen Li.shen@springer.com.

Please check our Lecture Notes in Mechanical Engineering at http://www.springer.com/series/11236 if you are interested in conference proceedings. To submit a proposal, please contact Leontina.dicecco@springer.com and Li.shen@springer.com.

More information about this series at http://www.springer.com/series/11693

Zhijie Liu · Jinkun Liu

PDE Modeling and Boundary Control for Flexible Mechanical System

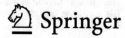

Zhijie Liu
School of Automation
and Electrical Engineering
University of Science
and Technology Beijing
Beijing, China

Jinkun Liu
School of Automation Science
and Electrical Engineering
Beihang University
Beijing, China

ISSN 2195-9862 ISSN 2195-9870 (electronic)
Springer Tracts in Mechanical Engineering
ISBN 978-981-15-2598-8 ISBN 978-981-15-2596-4 (eBook)
https://doi.org/10.1007/978-981-15-2596-4

Jointly published with Tsinghua University Press
The print edition is not for sale in China Mainland. Customers from China Mainland please order the print book from: Tsinghua University Press.

This Springer imprint is published by the registered company Springer Nature Singapore Pte Ltd.
The registered company address is: 152 Beach Road, #21-01/04 Gateway East, Singapore 189721, Singapore

Preface

In recent years, flexible mechanical systems possess many advantages over conventional rigid ones, such as lighter weight, lower cost and higher operation speed. These advantages motivate their wide application in the fields of spaceflight, aviation, machine, electron, chemical industry, biomedicine, etc. Modeling and vibration control of flexible mechanical systems have become a hot research topic, due to practical needs and theoretical challenges. Boundary control is an economical method of the distributed parameter systems (DPS), which is demonstrated by theoretical studies and industrial implementations. Since it requires few sensors and actuators, without decomposing the system into finite dimensional space. Therefore, in this book, modeling and boundary control problems are considered for several typical flexible mechanical systems based on partial differential equations (PDEs). This book is motivated by the fundamental issues including dynamic modeling and vibration control for different flexible mechanical systems. The main objectives of the book are to introduce the design method and MATLAB simulation of boundary control.

It is our goal to accomplish these objectives:

- Offer a catalog of modeling for different flexible mechanical systems based on partial differential equations.
- Provide advanced boundary controller design methods and their stability analysis methods.
- Offer simulation examples and MATLAB programs for some basic boundary control algorithms.

This book provides a thorough ground in PDE modeling and boundary controller design for flexible mechanical systems. Typical boundary controllers are verified using MATLAB. In this book, concrete cases are used to illustrate the successful application of the theory.

The book is structured as follows. The book starts with a brief introduction of boundary control for flexible mechanical systems in Chap. 1;

Chapter 2 presents the preliminaries, aerodynamic model and several lemmas for the subsequent development to simplify the dynamical modeling and further stability analysis for the beam structures.

In Chap. 3, we consider the modeling and control problem of a satellite with flexible solar panels. The panels with flexibility have been modeled as a distributed parameter system described by hybrid PDEs-ODEs. The control input has been proposed on the original PDE dynamics to suppress the vibrations of two panels. Then exponential stability has been proved by introducing a proper Lyapunov function.

In Chap. 4, we study vibration control to stabilize the flexible satellite described by a distributed parameter system modeled as PDEs with input constraint and external disturbance. In the controller design, an auxiliary system based on a smooth hyperbolic function and a Nussbaum function is adopted to handle the impact of the external disturbance and constrained input. The Lyapunov function is applied for control law design and stability analysis of the close-loop system.

In Chap. 5, a boundary control problem of a flexible aerial refueling hose in the presence of input disturbance is considered. To provide an accurate and concise representation of the hose's behavior, the flexible hose is modeled as a distributed parameter system described by PDEs. Boundary control is proposed based on the original PDEs to regulate the hose's vibration. A disturbance observer is designed to estimate the input disturbance.

In Chap. 6, a boundary control scheme is proposed based on backstepping method to regulate the hose's vibration. An auxiliary system based on a smooth hyperbolic function and a Nussbaum function is designed to handle the effect of the input saturation. With the Lyapunov's direct method, the closed-loop stability is proven through rigorous analysis without any simplification or discretization of the partial differential equation (PDE) dynamics.

In Chap. 7, we design a boundary control to stabilize a flexible hose modeled as a distributed parameter system (DPS) with input deadzone and output constraint. In the controller design, a radial basis function (RBF) neural network is used to handle the effect of the input deadzone, and a barrier Lyapunov function is employed to prevent constraint violation. The Lyapunov approach is applied for the stability analysis of the close-loop system.

In Chap. 8, by the extended Hamilton's principle, the flexible hose is modeled as a distributed parameter system described by PDEs. Then a boundary control scheme is proposed based on the original PDEs to regulate the hose's vibration and handle the effect of the input constraint in the presence of varying length varying speed and input constraint.

In Chap. 9, modeling and control design are presented for a three-dimensional flexible manipulator system with input disturbances. To provide an accurate and concise representation of the manipulator's dynamic behavior, the flexible manipulator is described by a distributed parameter system with a set of PDEs and ODEs. Boundary control laws with disturbance observers are designed to regulate orientation and suppress elastic vibrations simultaneously.

In Chap. 10, the conclusions are summarized and the future works are given.

In summary, this book covers the dynamical analysis, modeling and control design for three typical flexible mechanical systems. The book is primarily intended for researchers and engineers in the control system. It can also serve as a complementary reading on modeling and control of flexible mechanical systems at the postgraduate level.

Beijing, China   Zhijie Liu
September 2019   Jinkun Liu

Acknowledgements

We are very grateful for the opportunity to work with brilliant people who are generous with their time and friendship, through many discussions filled with creativity and inspiration. First and foremost, we would like to express our sincere appreciation to our co-workers who have contributed to the collaborative studies of this book.

We would also like to express our sincere appreciation to our colleagues who have contributed to the collaborative research. In particular, we would like to thank Miroslav Krstic, from the University of California, San Diego, US; Choon Ki Ahn, from the Korea University, South Korea; Keum-Shik Hong, from the Pusan National University, Korea; Huai-Ning Wu, from the Beihang University, China; Jun-Min Wang, from the Beijing Institute of Technology, China; We He, from the University of Science and Technology Beijing, China; Zhijia Zhao, from the Guangzhou University, and their research groups for their excellent research works, and helpful advice on our research. Special thanks go to Dongliang Sheng for his assistance and efforts in the process of publishing this book.

Appreciations must be made to Hongjun Yang, Yawei Peng, Tingting Jiang, Fangfei Cao, Ning Ji, Xueyan Xing, Shiqi Gao, Le Li, Xiaowei Zhang, Yongliang Yang, Yan Yang, Xiuyu He, Yuhua Song, Jiali Feng, Zhiji Han, Xuena Zhao, Jinyun Liang and Yonghao Ma for proofreading and providing numerous useful comments to improve the quality of the book.

Most importantly, we are deeply grateful to our families for their invaluable loves, supports and sacrifices over the years.

This work is supported by the National Key Research and Development Program of China under Grant 2019YFB1703603, the China Postdoctoral Science Foundation under Grant. 2019M660463, the Interdisciplinary Research Project for Young Teachers of USTB (Fundamental Research Funds for the Central Universities) under Grant FRF-IDRY-19-024, and Beijing Top Discipline for Artificial Intelligent Science and Engineering, University of Science and Technology Beijing.

Contents

About the Authors

Zhijie Liu received BS degree from China University of Mining and Technology (Beijing), Beijing in 2014, and the Ph.D. degree from Beihang University, Beijing in 2019. In 2017, he was a Research Assistant with the Department of Electrical Engineering, University of Notre Dame, for twelve months. He is currently an Associate Professor with the School of Automation and Electrical Engineering and Institute of Artificial Intelligence, University of Science and Technology Beijing, Beijing, China. He is a recipient of the IEEE SMC Beijing Capital Region Chapter Young Author Prize Award in 2019. He is the member of IEEE SMC Technical Committee on Autonomous Bionic Robotic Aircraft. His research interests include adaptive control, modeling and vibration control for flexible structures, and distributed parameter system.

Jinkun Liu was born on October 14, 1965, and he received BS, MS, and Ph.D. degrees from Northeastern University, Shenyang, China, in 1989, 1994, and 1997, respectively. He was a Postdoctoral Fellow in Zhejiang University from 1997 to 1999. He is currently a Full Professor in Beihang University, Beijing, P. R. China. His main research interest is adaptive boundary control for flexible manipulator. He has published more than 100 research papers and 8 books.

Chapter 1
Introduction

1.1 Background and Motivation

In recent decades, with the extensive application of flexible systems in the fields of biology, chemical engineering, medicine and aerospace and more and more challenges in theoretical research, dealing with vibration suppression of flexible systems has become an important research topic. Compared with traditional rigid mechanical systems, flexible mechanical systems overcome the disadvantages of low flexibility, high energy consumption, slow operating speed and limited operating space, and have the advantages of lighter material, high flexibility, low energy consumption, higher energy efficiency and higher operating speed. This promotes the application of flexible mechanical system in industrial field. There are a number of systems that can be described as flexible mechanical systems, such as telephone lines, conveyor belts, ocean pipelines, flexible manipulators, and solar panels. Here are three examples to illustrate their applications.

Example 1: Flexible manipulators.

The multi-degree-of-freedom flexible elephant trunk manipulator in Fig. 1.1 uses a motor to drive the flexible arm and uses the contraction or extension of artificial muscles to make the manipulator bend or straighten. This kind of manipulator is dexterous and gentle, and can be used in obstacle avoidance, service, medical treatment and other fields, which can greatly facilitate people's life.

Example 2: Flexible aerial refueling hose.

The aerial refueling hose system as shown in Fig. 1.2 has light and simple equipment. A drogue is a shuttlecock shape device installed at the end of hose, and it produces drag force for stability. A probe is a long pipe installed on the nose section of the receiver, and it couples with the drogue for refueling. The refueling process begins with the tanker deploying the hose with the drogue. The receiver pilot flies toward the drogue and couples with the probe. And the mechanical self-locking

© Tsinghua University Press 2020
Z. Liu and J. Liu, *PDE Modeling and Boundary Control for Flexible Mechanical System*, Springer Tracts in Mechanical Engineering,
https://doi.org/10.1007/978-981-15-2596-4_1

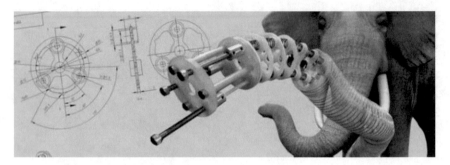

Fig. 1.1 Flexible manipulator with multiple degrees of freedom

(a) Aerial refueling process   (b) Refueling hose and cone sleeve

Fig. 1.2 Aerial refueling hose

mechanism on the drogue and probe is tightened, and the air docking operation is completed.

Example 3: Solar panels.

The futuristic Chinese space station, shown in Fig. 1.3, has two pairs of flexible solar panels. The solar panels, high-efficiency lithium ion battery, and other equipments constitute a power system of a space station, because of its light material. But because it is thin, long and flexible, it will produce strong vibration when it is affected by external disturbances such as cosmic wind or particle flow. If no effective measures are taken to suppress the vibration, the two pairs of "thin as cicada-wing" flexible structures will continue to withstand the vibration generated during the flight of the satellite, which may easily cause fatigue damage and lead to failure of expansion.

Therefore, flexible mechanical systems have very complex dynamical characteristics that are difficult to control compared with rigid mechanical systems. How to ensure the stability of flexible systems is one of the most important problems. Flexible mechanical systems are essentially infinite dimensional distributed parameter systems with complex dynamic characteristics. Previous studies on flexible mechanical systems are mostly based on lumped parameter models of ordinary differential equations (ODEs). Although the model description and controller design are simple

Fig. 1.3 Chinese space station

and easy to obtain, ODEs cannot accurately describe the dynamics of the system and may cause problems such as control overflow. The concept of distributed parameter system (DPS) originates from optimal control, which is compared with lumped parameter system (LPS). In mathematical terms, DPS uses partial differential equation (PDE) to describe how its state varies. In practice, many physical processes cannot be modeled with ODEs because their system states depend on two or more independent variables. The flexible mechanical system has two independent variables, space and time, so it is a distributed parameter model. Therefore, the model can be described by a series of infinite dimensional equations (describing the flexible body with PDE) and a series of finite dimensional equations (ODE equations describing the boundary conditions). Compared with LPS described by ordinary differential equation (ODE), the state space of DPS is an infinite dimensional space. Therefore, the flexible mechanical system based on PDE description brings some difficulties to theoretical analysis. At the same time, the vibration caused by the flexible characteristics of the flexible mechanical system brings some challenges to the extensive application of the flexible mechanical system. Therefore, it is urgent to find new control schemes to solve the control design problem of flexible mechanical systems based on PDEs, which is not only of great theoretical significance, but also of great engineering practical significance.

1.2 Survey on Modeling and Control of Flexible Mechanical System

1.2.1 Main Research Methods of Flexible Mechanical Systems Based on PDEs

Since the 1960s, many researchers all around the world have been studying the flexible mechanical systems based on DPS with the influence of the development of control theory and the actual needs of engineering, which proposing a lot of valuable research methods. To sum up, there are mainly operator semigroup method and Lyapunov direct method.

(1) Operator Semigroup Method

In the papers related to controller design of distributed parameter systems, functional method and operator semigroup theory are often used for stability analysis and the proof of PDE's well-posedness [12, 25, 48–51, 67, 78]. These distributed parameter systems are described by operator equations on infinite dimensional Hilbert or Banach Spaces, and then the stability analysis and the well-posedness of the solution can be obtained by the operator semigroup method on infinite dimensional space. In [13], the existence and uniqueness of the solution of the control system are proved by the infinite dimensional space theory. In [45], the authors use the operator semigroup theory to illustrate the asymptotic stability of the closed-loop system under the action of the controller. In [49], the stability of the system is studied by applying semigroup theory. In [21], the strong stability of one-dimensional wave equation under the action of controller is proved by using semigroup theory. The stability of a second order PDE system with unmatched control and observation in Hilbert space is studied in [23]. In [22], the authors propose an unmatched boundary control method to stabilize two connected soft ropes, and then obtain the exponential stability of the closed-loop system by using the semigroup theory. In [78], an observer of exponential convergence for distributed parameter systems is developed, and the stability and well-posedness of the system are proved by using semigroup theory. For engineering designers, this analysis method requires more profound mathematical knowledge, which to some extent limits the application in practical engineering.

(2) Lyapunov Direct Method

Lyapunov direct method in the ODE system can analyze the stability of the system by using the properties of derivation of $V(x)$ without solving the system state equation or locally linearizing the nonlinear system. Lyapunov function $V(x)$ ($x \in \mathbb{R}$ represents the state of the system) represents the energy of the system, and $\dot{V}(x)$ represents the trend of energy. Lyapunov direct method is widely applied to the analysis and synthesis problems of ODE systems because of its advantages such as simple and intuitive, wide application range and easy accepted by engineering designers. In order to overcome the shortcoming that the operator semigroup method needs deep theory and cannot be accepted by engineers easily, some scholars try to establish a system analysis method based on Lyapunov direct method directly from the original DPS system model. Many scholars have put forward the boundary control method of flexible system based on Lyapunov direct method [28, 31, 36, 37, 69, 81, 82, 88–91]. In [36], He Wei established a distributed parameter model for the flexible wings of a robotic aircraft and designed a boundary control for the PDE model based on direct method of Lyapunov. In [59], a robust adaptive control method was proposed to suppress the vibration of a stretched string. In [60], Rahn presents a boundary control strategy with exponential convergence for a flexible cable gantry crane. In [11], an adaptive controller is developed by Lyapunov method for two kinds of distributed parameter systems (string and noise), and an analysis and controller design method for general nonlinear infinite dimensional system is presented.

Compared with operator semigroup method, Lyapunov direct method requires less background knowledge for distributed parameter system, so it is easier to design the controller and analyze systems' stability. In addition, Lyapunov direct method provides a simple and direct method for PDEs by using easy-to-understand mathematical tools, such as algebraic integration inequalities and integration by parts.

1.2.2 Control Design of Flexible Mechanical Systems

As the dynamic equations of flexible mechanical systems based on PDEs are infinite dimensional distributed parameter systems, many control strategies of traditional rigid system cannot be directly used to the flexible mechanical systems. At present, there are three widely used control methods based on distributed parameter system: modal control, distributed control and boundary control. Modal control method is based on truncated models when design control strategies. The general reduction methods are finite element method, Galerkin method and Assumed mode method.

(1) Finite Element Method, FEM

The finite element method is to divide infinite degrees of freedom on flexible mechanical system into finite connected elements to simplify the problem. These elements are connected by nodes, and the displacement of each element is described by the interpolation function of node displacement. The selection of interpolation function is relatively simple, because it is essentially a hypothetical mode, not for the whole structure and each substructure. However, this method requires a large amount of computation, so is only suitable for computer operation. In [18], the authors give a nonlinear dynamic model of a single-link flexible manipulator by using finite element method and Lagrange motion equation. In [52, 53, 77], Martins and Tokhi et al. use finite element method to give a dynamic model of a single-link flexible robotic manipulator and modify the model through experiments. Rosado et al. establish the dynamic model of the flexible mechanical system by using the finite element method combined with Newton–Euler equation [17, 76].

(2) Galerkin Method, GM

Galerkin method is a numerical analysis method proposed by Russian mathematician Galerkin in the early 20th century. The basic principle of Galerkin method is that the original functional can be approached piecewise by constructing a set of bases in the N dimensional space, so that the optimal solution is only needed in the current space (not necessarily making the original equation strictly true, but only making it approximately equal). By changing the dimension of the space, the accuracy can be improved continuously. Then the problem of solving the original variational problem was transformed to the problem of finding the coefficient of the basis function. However, as a trial function selection form of the method of weighted residual, what Galerkin method obtained was only an approximate solution in the original solution domain. In [41], Kim et al. establish a dynamic model of flexible tethered satellite systems by Newton–Euler method, compare and analyze the interaction between transverse and longitudinal tension and motion in the recovery process of system by using three-order Galerkin model and classical bead model.

(3) Assumed Mode Method, AMM

The assumed mode method is to use the mode truncation technique to acquire the eigenvalue of free vibration, then obtain the dynamic mode of each substructure, and finally obtain the modal of the whole system. Based on the Rayleigh–Ritz method, the solution of the system is expanded into an infinite series, and then the sum of the first n modes is used to approximate the modes of the whole system. The method can be further divided into constrained and unconstrained modal methods. In [3, 68], a dynamic model of a single-link flexible arm is given by using the assumed mode method and Hamilton principle. In [9], by using assumed mode method and Lagrange motion equation, the authors solve the characteristic root problem of matrix equation of a multi-degree-of-freedom stiffened plate system and obtain the natural frequency and mode of the system. Rakhsha et al. in [61] establish a single-link flexible robot model by Newton–Euler equation and constrained mode method. Sun et al. in [70–72] establish a dynamic model for the manipulator's cooperative control of flexible load, then design an observer based on the assumed mode method, and finally prove the stability of the closed-loop system. Although these three methods simplify the design and analysis of the control system, ignoring the high order modes of the flexible system may cause the instability of the actual system, that is, the spillover problem of observation or control. In [1, 57], the authors study the spillover problem when the controller is designed on truncated model based on few modes. In order to obtain more accurate control performance, higher order system modes are needed, which will increase the order of the controller. This increases the difficulty of actual control system design and implementation. In order to overcome the shortcomings of truncated model-based mode control, distributed control and boundary control can achieve very good control performance. Distributed control [2, 80, 84, 85] requires more actuators and sensors, which makes distributed control difficult to implement. Compared with distributed control, boundary control is a more effective control method. It does not need to reduce the order of the system to a finite dimensional model. This control method only needs to use actuators and sensors at the boundary of the system. Moreover, boundary control can be combined with other classical control methods to achieve the desired control performance.

1.2.3   Typical Flexible Mechanical Systems

In practical control problems, common PDEs include Euler–Bernoulli equations [20, 35], string equations [29, 83], etc. The system models involved in this book include flexible satellite systems and flexible manipulator systems based on Euler–Bernoulli equations (including models in two-dimensional and three-dimensional space), and flexible aerial refueling systems based on string equations. The following is a systematic introduction of modeling and control methods for these systems.

(1) Flexible Satellite Systems

Satellites have gained considerable interest in the past decades as a result of applications in communication, remote sensing, etc. [92–94]. Under a complex environment, space mission constraints have pushed demands such as lighter weight structure, limitation of mass, low energy consumption, and reduced launch cost. A number of satellites with a rigid hub and long flexible solar panels are used in space missions. Recently, a number of methods have been developed for the control of flexible satellites, including adaptive fuzzy sliding mode control [95], H-infinity control [98], variable structure attitude control [96, 97] and so on.

Due to the flexible property of solar panels, the deflection of the flexible panels has a significant influence on the dynamics and control performance of the satellites. Therefore, vibration suppression is an important research topic related to flexible spacecraft.

(2) Flexible Aerial Refueling Systems

Aerial is the main method to increase the range in the aviation field. It is mainly used to improve the endurance and combat radius of fighter aircraft in the military field, and the safety and economic performance in the civil field. At present, the main refueling methods include flying truss type, referred to as hard tube type and cone-sleeve type, referred to as hose type. Compared with the hard tube aerial refueling, the hose type has the advantages of light weight, low cost, and the ability to refueling multiple oil receivers at the same time without the need for additional refueling operators, thus gaining more extensive applications. However, due to the flexible property of the hose, docking is difficult. Therefore, the vibration suppression of hose has received a lot of attention and research [65, 75, 79].

Modeling and controller design of hose-cone sleeve system in the docking process of tanker and receiver are presented in [66]. In [63, 64], the authors first proposed and studied the modeling and simulation of controllable cone sleeve. The above mentioned research on tanker hose-cone sleeve is based on finite element method and centralized mass method. As mentioned above, these two methods are both "order reduction before design" methods, which cannot avoid the defects of controller design based on truncation model.

The tanker hose is DPS in essence, and the controller should be designed on the basis of the original PDE model, such as boundary control. As far as we know, there is few relevant literatures to study the controller design of flexible hose of aerial refueling tanker based on PDE model. But the research of like second-order string (soft rope or cable) equations have made great progress. However, the modeling of the oil hose of aerial tanker is different from the several string models mentioned above, because the hose is affected by horizontal motion and gravity. Different from the axial movement in [86, 91], the movement of flexible hose have a certain Angle, which makes the model more complicated. If the time-varying length of flexible pipeline is considered, the modeling and control of tanker hose will be more difficult. Therefore, the research on the oil hose of aerial refueling tanker based on PDE model needs further research. At the same time, the limited control input, the control performance design of the system and the control design in the presence of system's uncertainty and external interference also need to be solved.

(3) Flexible Manipulator Systems

The control tasks of flexible manipulator system can be divided into four categories: vibration suppression [6, 24, 44], position control [4, 19, 40, 87], force and position hybrid control [16, 54, 55] and the cooperation control [8, 39, 56] of flexible manipulators.

Position control and vibration suppression are the basic problems in the research of control of flexible manipulator systems, where there are mainly the following control methods: PD control [19, 40], feedback linearization [4, 10], robust control [7], adaptive control [87], intelligent control [26, 38, 73] and optimal control [5, 42, 46]. Although some research results have been achieved on the boundary control method, further research is still needed, and the following problems remain to be solved: The current results are mainly based on the assumption that the control input is ideal, but the actuators in the actual system are limited. How to design a control strategy with constrained control input to obtain satisfactory control performance. As flexible manipulator system is mainly used in aviation and medical and other occasions of high precision requirements, how to ensure that the control performance of the system meet the expected indicators. How to design control schemes when the system has uncertainty and external disturbances and so on.

(4) Flexible Manipulators in 3D space

Similar to flexible manipulators in two-dimensional plane, the control research of traditional flexible manipulators in three-dimensional space is basically based on truncation models, which inevitably bring problems such as control overflow and high dimension of controller. Therefore, we need to explore the control methods based on the original PDEs. The flexible manipulator in the three-dimensional space is also composed of a central rigid body and a flexible beam, in which the flexible beam is simplified to Euler–Bernoulli beam. Although there have been many studies on flexible mechanical systems based on PDEs, such as distributed control and boundary control mentioned above, these researches are all based on the flexible mechanical system of plane motion, ignoring the coupling effect between deformation in different directions, which will damage the control performance of the system. Therefore, it is necessary to consider the coupling action between all directions in the controller design of flexible mechanical system. At present, many scholars have started to study the control of 3D flexible mechanical systems. Nguyen et al. in [58] propose a boundary controller for 3D marine riser and proved the global stability of the closed-loop system based on the original PDE model with Lyapunov direct method. Do et al. in [14] design a boundary control strategy for a fixed-length marine riser based on Lyapunov direct method and backstepping method, and analyzed the existence and uniqueness of the solution of the system. Wei He et al. study the modeling and boundary control of 3D Euler–Bernoulli beam and 3D flexible rope in [33, 35], and proved the stability of the system by Lyapunov direct method. Although researches on the modeling and control of flexible mechanical system in three-dimensional space have made great progress, the modeling and control of flexible manipulator in three-dimensional space based on PDE model still need to be solved. It is necessary to consider the direction tracking and vibration suppression of 3D flexible manipulator when it moves in space. In three dimensional space, there are strong coupling of flexible deformation in three directions and the coupling of deformation and motion attitude, which make modeling and controller design more difficult. At the same time, considering that the uncertainty of the model and the phenomenon of external interference exist in a large number of actual projects, it is of great theoretical and practical significance to solve the problem of 3D modeling and control of flexible manipulators.

1.2.4 Engineering Design of Flexible Mechanical Systems

In the process of practical applications, the performance of control systems cannot be based on ideal control inputs or only meet the theoretical stability requirements. A series of engineering problems need to be considered, such as control input constraints, actuator failures, external interference, system output constraints and so on. This book mainly considers two important problems in flexible mechanical systems: control input constraints and output constraints.

(1) Control input constraints

In order to achieve the desired control performance of the control systems, there may be cases where the control input needs to be very large. In the theoretical design, we believe that the actuator can output any desired control signal to meet the control requirements. However, in the practical application processes, the controllers have the upper limit of input because the controller's physical structure or energy supply is limited to some extent. When even the maximum output control signals of the actuators cannot meet the ideal inputs, the systems may be unstable. In addition, keeping the actuators in a saturated state will not only reduce the life of the actuators, but also seriously damage the actuators causing safety problems. Therefore, it is very important to study how to meet the expected control performance when the control inputs are limited and to realize the stability of the systems. At present, the problems of limited inputs for flexible mechanical systems based on distributed parameter systems are very few. He in [32, 34] design boundary control and vibration control based on saturation of control input for a flexible rope system. The boundary control strategies with input constraints for flexible mechanical systems, especially for systems with more complex models, need to be further studied.

(2) Control output constraints

For flexible mechanical systems, many tasks not only require the realization of the most basic control needs, such as position and attitude tracking and vibration suppression, but also need to consider the control performance, that is, the output of the system. For example, position and attitude tracking errors or elastic vibrations deformation amplitude need to meet certain constraints. Therefore, in practical applications, the output limitations of the systems are very common, especially in the aspects of safety regulations. The violation of the restriction may lead to the performance degradation or instability of the systems, or even cause serious damage to the systems and endanger the security. For example, when flexible arms are used in medical operations, excessive vibration may threaten the safety of patients. Therefore, it is very important to study the output constraint of the system. In [15], Dubljevic designs a control method for a class of hyperbolic PDE systems under input and state constraints. Lee and Ren et al. in [62, 74] design a novel Barrier Lyapunov Function (BLF), which solve the control problem of a class of nonlinear systems with limited output. He et al. [30] extend the method and apply it to a flexible manipulator, which solve the problem of system with output constraint and make the output of the system meet the given boundary. Considering that the barrier Lyapunov function based control can only deal with constant boundary constraint, a prescribed performance function is proposed in [43]. Using a prescribed error transformation function, a prescribed performance can be guaranteed, which means that the tracking performance of the transient and steady state error can be regulated. However, this method is designed for ODE systems, and the control of flexible mechanical systems based on PDEs needs further study.

1.3 Outline of This Book

The general objectives of the book are to develop constructive and systematic methods of modeling and designing control for flexible mechanical systems. By investigating the characteristics of several different models, control strategies are proposed to achieve the performance for the concerned systems. The book starts with a brief introduction of modeling and control techniques for flexible mechanical systems in this chapter.

Chapter 2 presents the preliminaries, aerodynamic model and several lemmas for the subsequent development to simplify the dynamical modeling and further stability analysis for the beam structures.

In Chap. 3, we consider the modeling and control problem of a satellite with flexible solar panels. The panels with flexibility have been modeled as a distributed parameter system described by hybrid PDEs-ODEs. The control input has been proposed on the original PDE dynamics to suppress the vibrations of two panels. Then exponential stability has been proved by introducing a proper Lyapunov function.

In Chap. 4, we study vibration control to stabilize the flexible satellite described by a distributed parameter system modeled as PDEs with input constraint and external disturbance. In the controller design, an auxiliary system based on a smooth hyperbolic function and a Nussbaum function is adopted to handle the impact of the external disturbance and constrained input. The Lyapunov function is applied for control law design and stability analysis of the close-loop system.

In Chap. 5, a boundary control problem of a flexible aerial refueling hose in the presence of input disturbance is considered. To provide an accurate and concise representation of the hose's behavior, the flexible hose is modeled as a distributed parameter system described by PDEs. Boundary control is proposed based on the original PDEs to regulate the hose's vibration. A disturbance observer is designed to estimate the input disturbance.

In Chap. 6, a boundary control scheme is proposed based on backstepping method to regulate the hose's vibration. An auxiliary system based on a smooth hyperbolic function and a Nussbaum function is designed to handle the effect of the input saturation. With the Lyapunov's direct method, the closed-loop stability is proven through rigorous analysis without any simplification or discretization of the partial differential equation (PDE) dynamics.

In Chap. 7, we design a boundary control to stabilize a flexible hose modeled as a distributed parameter system (DPS) with input deadzone and output constraint. In the controller design, a radial basis function (RBF) neural network is used to handle the effect of the input deadzone, and a barrier Lyapunov function is employed to prevent constraint violation. The Lyapunov approach is applied for the stability analysis of the close-loop system.

In Chap. 8, by the extended Hamilton's principle, the flexible hose is modeled as a distributed parameter system described by PDEs. Then a boundary control scheme is proposed based on the original PDEs to regulate the hose's vibration and handle the effect of the input constraint in the presence of varying length varying speed and input constraint.

In Chap. 9, modeling and control design are presented for a three-dimensional flexible manip-ulator system with input disturbances. To provide an accurate and concise representation of the manipulator's dynamic behavior, the flexible manipulator is described by a distributed param-eter system with a set of PDEs and ODEs. Boundary control laws with disturbance observers are designed to regulate orientation and suppress elastic vibrations simultaneously.

In Chap. 10, the conclusions are summarized and the future works are given.

References

1. Balas MJ (1978) Active control of flexible systems. J Optim Theory Appl 25(3):415–436
2. Bamieh B, Paganini F, Dahleh MA (2002) Distributed control of spatially invariant systems. IEEE Trans Autom Control 47(7):1091–1107
3. Barbieri E, Ozguner U (1988) Unconstrained and constrained mode expansions for a flexible slewing link. J Dyn Syst Meas Control 110(4):416–421
4. Cai G-P, Lim CW (2006) Active control of a flexible hub-beam system using optimal tracking control method. Int J Mech Sci 48(10):1150–1162
5. Cai G-P, Lim CW (2006) Optimal tracking control of a flexible hub-beam system with time delay. Multibody Syst Dyn 16(4):331–350
6. Canbolat H, Dawson D, Rahn C, Vedagarbha P (1998) Boundary control of a cantilevered flexible beam with point-mass dynamics at the free end. Mechatronics 8(2):163–186
7. Caracciolo R, Richiedei D, Trevisani A, Zanotto V (2005) Robust mixed-norm position and vibration control of flexible link mechanisms. Mechatronics 15(7):767–791
8. Chiu C-S, Lian K-Y, Tsu-Cheng W (2004) Robust adaptive motion/force tracking control design for uncertain constrained robot manipulators. Automatica 40(12):2111–2119
9. Cho D, Vladimir N, Choi T (2015) Natural vibration analysis of stiffened panels with arbitrary edge constraints using the assumed mode method. Proc Inst Mech Eng Part M: J Eng Marit Environ 229(4):340–349
10. De Luca A, Lanari L (1995) Robots with elastic joints are linearizable via dynamic feedback. In Proceedings of the 34th IEEE Conference on Decision and Control, 1995, vol 4. IEEE, pp 3895–3897
11. De Queiroz MS, Rahn CD (2002) Boundary control of vibration and noise in distributed parameter systems: an overview. Mech Syst Signal Process 16(1):19–38
12. Demetriou MA (2004) Natural second-order observers for second-order distributed parameter systems. Syst Control Lett 51(3):225–234
13. Do KD, Pan J (2008) Boundary control of transverse motion of marine risers with actuator dynamics. J Sound Vib 318(4):768–791
14. Do KD, Pan J (2009) Boundary control of three-dimensional inextensible marine risers. J Sound Vib 327(3–5):299–321
15. Dubljevic S, El-Farra NH, Mhaskar P, Christofides PD (2006) Predictive control of parabolic pdes with state and control constraints. Int J Robust Nonlinear Control: IFAC-Affil J 16(16):749–772
16. Endo T, Matsuno F (2004) Dynamics based force control of one-link flexible arm. In SICE 2004 Annual Conference, vol 3. IEEE, pp 2736–2741
17. Gamarra-Rosado VO, Yuhara EAO (1999) Dynamic modeling and simulation of a flexible robotic manipulator. Robotica 17(5):523–528
18. Ge SS, Lee TH, Zhu G (1997) A nonlinear feedback controller for a single-link flexible manipulator based on a finite element model. J Robot Syst 14(3):165–178
19. Ge SS, Lee TH, Zhu G (1998) Improving regulation of a single-link flexible manipulator with strain feedback. IEEE Trans Robot Autom 14(1):179–185

20. Ge SS, Zhang S, He W (2011) Vibration control of an euler-bernoulli beam under unknown spatiotemporally varying disturbance. Int J Control 84(5):947–960
21. Guo BZ, Guo W (2009) The strong stabilization of a one-dimensional wave equation by non-collocated dynamic boundary feedback control. Automatica 45(3):790–797
22. Guo BZ, Jin FF (2010) Arbitrary decay rate for two connected strings with joint anti-damping by boundary output feedback. Automatica 46(7):1203–1209
23. Guo BZ, Shao ZC (2009) Stabilization of an abstract second order system with application to wave equations under non-collocated control and observations. Syst Control Lett 58(5):334–341
24. Guo B-Z, Wang J-M (2005) The well-posedness and stability of a beam equation with conjugate variables assigned at the same boundary point. IEEE Trans Autom Control 50(12):2087–2093
25. Guo BZ, Xu CZ (2007) The stabilization of a one-dimensional wave equation by boundary feedback with noncollocated observation. IEEE Trans Autom Control 52(2):371–377
26. Gutierrez LB, Lewis FL, Lowe JA (1998) Implementation of a neural network tracking controller for a single flexible link: comparison with pd and pid controllers. IEEE Trans Ind Electron 45(2):307–318
27. He W, Zhang S, Ge SS (2013) Boundary control of a flexible riser with the application to marine installation. IEEE Trans Ind Electron 60(12):5802–5810
28. He W, Yan Z, Sun C, Chen Y (2017) Adaptive neural network control of a flapping wing micro aerial vehicle with disturbance observer. IEEE Trans Cybern 47(10):3452–3465
29. He W, Ge SS (2012) Robust adaptive boundary control of a vibrating string under unknown time-varying disturbance. IEEE Trans Control Syst Technol 20(1):48–58
30. He W, Ge SS (2015) Vibration control of a flexible beam with output constraint. IEEE Trans Ind Electron 62(8):5023–5030
31. He W, Ge SS (2016) Cooperative control of a nonuniform gantry crane with constrained tension. Automatica 66:146–154
32. He W, He X, Ge SS (2015) Boundary output feedback control of a flexible string system with input saturation. Nonlinear Dyn 80(1–2):871–888
33. He W, Qin H, Liu J-K (2015) Modelling and vibration control for a flexible string system in three-dimensional space. IET Control Theory Appl 9(16):2387–2394
34. He W, Sun C, Ge SS (2014) Vibration control design for a flexible string with input saturation. In 2014 11th World Congress on Intelligent Control and Automation (WCICA). IEEE, pp 885–890
35. He W, Yang C, Zhu J, Liu J-K, He X (2017) Active vibration control of a nonlinear three-dimensional euler-bernoulli beam. J Vib Control 23(19):3196–3215
36. He W, Zhang S (2017) Control design for nonlinear flexible wings of a robotic aircraft. IEEE Trans Control Syst Technol 25(1):351–357
37. He X, Wei H, Jing S, Sun C (2017) Boundary vibration control of variable length crane systems in two dimensional space with output constraints. IEEE/ASME Trans Mechatron 22(5):1952–1962
38. Jnifene A, Andrews W (2005) Experimental study on active vibration control of a single-link flexible manipulator using tools of fuzzy logic and neural networks. IEEE Trans Instrum Meas 54(3):1200–1208
39. Karayiannidis Y, Rovithakis G, Doulgeri Z (2007) Force/position tracking for a robotic manipulator in compliant contact with a surface using neuro-adaptive control. Automatica 43(7):1281–1288
40. Kelly R, Ortega R, Ailon A, Loria A (1994) Global regulation of flexible joint robots using approximate differentiation. IEEE Trans Autom Control 39(6):1222–1224
41. Kim E, Vadali SR (1995) Modeling issues related to retrieval of flexible tethered satellite systems. J Guid Control Dyn 18(5):1169–1176
42. Korayem MH, Nikoobin A, Azimirad V (2009) Trajectory optimization of flexible link manipulators in point-to-point motion. Robotica 27(6):825–840

43. Kostarigka AK, Doulgeri Z, Rovithakis GA (2013) Prescribed performance tracking for flexible joint robots with unknown dynamics and variable elasticity. Automatica 49(5):1137–1147
44. Krstic M, Siranosian AA, Balogh A, Guo BZ (2007) Control of strings and flexible beams by backstepping boundary control. In American Control Conference, 2007. ACC'07. IEEE, pp 882–887
45. Kyung-Jinn Y, Keum-Shik H, Fumitoshi M (2004) Robust adaptive boundary control of an axially moving string under a spatiotemporally varying tension. J Sound Vib 273(4):1007–1029
46. Lahdhiri T, ElMaraghy HA (1998) ElMaraghy. Optimal nonlinear position tracking control of a two-link flexible-joint robot manipulator. In *Experimental Robotics V*, Experimental Robotics V. Springer, pp 502–514
47. Liu J, He W (2018) Distributed parameter modeling and boundary control of flexible manipulators. Springer Singapore, Singapore
48. Luo ZH (1993) Direct strain feedback control of flexible robot arms: new theoretical and experimental results. IEEE Trans Autom Control 38(11):1610–1622
49. Luo ZH, Guo BZ, Morgül Ö (2012) Stability and stabilization of infinite dimensional systems with applications. Springer Science & Business Media
50. Luo ZH, Guo B (1995) Further theoretical results on direct strain feedback control of flexible robot arms. IEEE Trans Autom Control 40(4):747–751
51. Luo ZH, Kitamura N, Guo BZ (1995) Shear force feedback control of flexible robot arms. IEEE Trans Robot Autom 11(5):760–765
52. Martins JM, Mohamed Z, Tokhi MO, Sa Da Costa J, Botto MA (2003) Approaches for dynamic modelling of flexible manipulator systems. IEE Proc-Control Theory Appl 150(4):401–411
53. Martins J, Botto MA, Da Costa JS (2002) Modeling of flexible beams for robotic manipulators. Multibody Syst Dyn 7(1):79–100
54. Matsuno F, Asano T, Sakawa Y (1994) Modeling and quasi-static hybrid position/force control of constrained planar two-link flexible manipulators. IEEE Trans Robot Autom 10(3):287–297
55. Matsuno F, Kasai S (1998) Modeling and robust force control of constrained one-link flexible arms. J Robot Syst 15(8):447–464
56. Harris McClamroch N, Wang D (1988) Feedback stabilization and tracking of constrained robots. IEEE Trans Autom Control 33(5):419–426
57. Meirovitch L, Baruh H (1983) On the problem of observation spillover in self-adjoint distributed-parameter systems. J Optim Theory Appl 39(2):269–291
58. Nguyen TL, Do KD, Pan J (2013) Boundary control of coupled nonlinear three dimensional marine risers. J Mar Sci Appl 12(1):72–88
59. Zhihua Q (2001) Robust and adaptive boundary control of a stretched string on a moving transporter. IEEE Trans Autom Control 46(3):470–476
60. Rahn CD, Zhang F, Joshi S, Dawson DM (1999) Asymptotically stabilizing angle feedback for a flexible cable gantry crane. J Dyn Syst Meas Control 121(3):563–566
61. Rakhsha F, Goldenberg A (1985) Dynamics modelling of a single-link flexible robot. In Robotics and Automation. Proceedings. 1985 IEEE International Conference on, vol 2.IEEE, pp 984–989
62. Ren B, Ge SS, Tee KP, Lee TH (2010) Adaptive neural control for output feedback nonlinear systems using a barrier lyapunov function. IEEE Trans Neural Netw 21(8):1339–1345
63. Ro K, Kamman JW (2010) Modeling and simulation of hose-paradrogue aerial refueling systems. J Guid Control Dyn 33(1):53–63
64. Ro K, Kuk T, Kamman J (2010) Active control of aerial refueling hose-drogue systems. In AIAA Guidance, Navigation, and Control Conference. p 8400
65. Ro K, Kuk T, Kamman J (2011) Design, test and evaluation of an actively stabilized drogue refueling system. In *Infotech@ Aerospace 2011*, Infotech@ Aerospace 2011, p 1423
66. Ro K, Kuk T, Kamman JW (2011) Dynamics and control of hose-drogue refueling systems during coupling. J Guid Control Dyn 34(6):1694–1708

67. Sakawa Y, Luo ZH (2002) Dynamics and control of bending and torsional vibrations of flexible beams. IEEE Trans Autom Control 34(9):970–977
68. Singh TR (1991) Dynamics and control of flexible arm robots. Thesis Waterloo University
69. Su L, Wang J-M, Krstic M (2018) Boundary feedback stabilization of a class of coupled hyperbolic equations with nonlocal terms. IEEE Trans Autom Control 63(8):2633–2640
70. Sun D, Liu YH (2001) Position and force tracking of a two-manipulator system manipulating a flexible beam. J Robot Syst 18(4):197–212
71. Sun D, Liu Y-H (2001) Position and force tracking of a two-manipulator system manipulating a flexible beam. J Robot Syst 18(4):197–212
72. Sun D, Mills JK, Liu Y (1998) Hybrid position and force control of two industrial robots manipulating a flexible sheet: Theory and experiment. In Proceedings. 1998 IEEE International Conference on Robotics and Automation, 1998, vol 2. IEEE, pp 1835–1840
73. Tang Y, Sun F, Sun Z (2006) Neural network control of flexible-link manipulators using sliding mode. Neurocomputing 70(1–3):288–295
74. Tee KP, Ge SS, Tay EH (2009) Barrier lyapunov functions for the control of output-constrained nonlinear systems. Automat 45(4):918–927
75. Thomas PR, Bhandari U, Bullock S, Richardson TS, Du Bois JL (2014) Advances in air to air refuelling. Prog Aerosp Sci 71:14–35
76. Tian Q, Zhang Y, Chen L, Yang JJ (2010) Simulation of planar flexible multibody systems with clearance and lubricated revolute joints. Nonlinear Dyn 60(4):489–511
77. Osman Tokhi M, Mohamed Z, Hasan Shaheed M (2001) Dynamic characterisation of a flexible manipulator system. Robotica 19(5):571–580
78. Nguyen DT (2009) Boundary output feedback of second-order distributed parameter systems. Syst Control Lett 58(7):519–528
79. Williamson WR, Reed E, Glenn GJ, Stecko SM, Musgrave J, Takacs JM (2010) Controllable drogue for automated aerial refueling. J Aircr 47(2):515–527
80. Wu F (2003) Distributed control for interconnected linear parameter-dependent systems. IEE Proc-Control Theory Appl 150(5):518–527
81. Wu HN, Feng S (2017) Guaranteed-cost vinite-time fuzzy control for temperature-constrained nonlinear coupled beat-ode systems. IEEE Trans Syst Man Cybern Syst 47(8):1919–1930
82. Wu HN, Wang HD (2017) Distributed consensus observers-vased h_∞ control of dissipative pde systems using sensor networks. IEEE Trans Control Netw Syst 2(2):112–121
83. Wu HN, Wang J-W, Li H-X (2014) Fuzzy boundary control design for a class of nonlinear parabolic distributed parameter systems. IEEE Trans Fuzzy Syst 22(3):642–652
84. Yang H, Liu J (2016) Distributed piezoelectric vibration control for a flexible-link manipulator based on an observer in the form of partial differential equations. J Sound Vib 363:77–96
85. Yang HJ, Liu JK, He W (2018) Distributed disturbance-observer-based vibration control for a flexible-link manipulator with output constraints. Sci China Technol Sci 61(10):1528–1536
86. Yang K-J, Hong K-S, Matsuno F (2005) Robust boundary control of an axially moving string by using a pr transfer function. IEEE Trans Autom Control 50(12):2053–2058
87. Zhang L, Liu J (2013) Adaptive boundary control for flexible two-link manipulator based on partial differential equation dynamic model. IET Control Theory Appl 7(1):43–51
88. Zhang S, Dong Y, Ouyang Y, Yin Z, Peng K (2018) Adaptive neural control for robotic manipulators with output constraints and uncertainties. IEEE Trans Veural Vetworks Learn Syst 99:1–11
89. Zhang YL, Wang JM (2017) Exact controllability of a micro beam with boundary bending moment. Int J Control 1–9
90. Zhao Z, Liu Y, Luo F (2017) Output feedback boundary control of an axially moving system with input saturation constraint. ISA Trans 68:22–32
91. Zhao Z, Liu Y, Fang G, Yun F (2017) Vibration control and boundary tension constraint of an axially moving string system. Nonlinear Dyn 89(1):1–10
92. Sabatini M, Palmerini GB, Leonangeli N, Gasbarri P (2014) Analysis and experiments for delay compensation in attitude control of flexible spacecraft. Acta Astronaut 104(1):276–292

93. Meng D, Wang X, Wenfu X, Liang B (2017) Space robots with flexible appendages: dynamic modeling, coupling measurement, and vibration suppression. J Sound Vib 396:30–50
94. Wenfu X, Meng D, Chen Y, Qian H, Yangsheng X (2014) Dynamics modeling and analysis of a flexible-base space robot for capturing large flexible spacecraft. Multibody Syst Dyn 32(3):357–401
95. Guan P, Liu X-J, Liu J-Z (2005) Adaptive fuzzy sliding mode control for flexible satellite. Eng Appl Artif Intell 18(4):451–459
96. Qinglei H, Ma G (2005) Variable structure control and active vibration suppression of flexible spacecraft during attitude maneuver. Aerosp Sci Technol 9(4):307–317
97. Qinglei H, Ma G (2008) Adaptive variable structure controller for spacecraft vibration reduction. IEEE Trans Aerosp Electron Syst 44(3):861–876
98. Cubillos XC, de Souza GLC (2009) Using of h infinity control method in attitude control system of rigid-flexible satellite. Math Probl Eng 2009
99. Yang K-J, Hong K-S, Matsuno F (2005) Energy-based control of axially translating beams: varying tension, varying speed, and disturbance adaptation. IEEE Trans Control Syst Technol 13(6):1045–1054

Chapter 2
Mathematical Preliminaries

In this chapter, we briefly introduce some important theoretical basics such as lemma and aerodynamic model involved in this book.

2.1 Basic PDE Models

PDE is the equation that reflects the constraint relationship between time and space variables, and models in many domains can be described by PDE. Here we introduce two kinds of partial differential equations that are used in this book and are most widely used in engineering field.

(1) Wave Equation

$$w_{tt}(x, t) = w_{xx}(x, t) \tag{2.1}$$

Equation (2.1) is the simplest and common second-order PDE, which can describe a variety of common oscillation phenomena, such as the vibration of the flexible rope. In general, to solve the solution under specific conditions of PDE, we need to determine the condition of the solution, which is called as definite condition. There are two kinds of definite conditions: boundary conditions and initial conditions. In practical application problems, both of the conditions of the equation are usually given, and the boundary conditions have corresponding physical meaning. For example, the boundary condition of (2.1) is set as

$$w(1, t) = 0, w_x(0, t) = 0 \tag{2.2}$$

which means that the end of the rope is fixed at $x = 1$, free at $x = 0$, and the slope of 0 means that no external force is applied at $x = 0$.

(2) Beam Equation

$$w_{tt}(x, t) = w_{xxxx}(x, t) \tag{2.3}$$

© Tsinghua University Press 2020
Z. Liu and J. Liu, *PDE Modeling and Boundary Control for Flexible Mechanical System*, Springer Tracts in Mechanical Engineering,
https://doi.org/10.1007/978-981-15-2596-4_2

Equation (2.3) is the Euler–Bernoulli beam equation, which is widely used to describe the vibration of flexible body, such as flexible manipulator and solar panels, etc. Its boundary condition can be denoted by

$$w_x(0, t) = 0, \ w_{xxx}(0, t) = 0 \tag{2.4}$$
$$w(1, t) = 0, \ w_{xx}(l, t) = 0 \tag{2.5}$$

The most obvious difference between Eqs. (2.1) and (2.3) is that the wave equation is the second-order spatial derivative PDE, and the beam equation is the fourth-order spatial derivative PDE. Accordingly, the wave equation has one boundary condition for each endpoint and the beam equation has two boundary conditions for each endpoint. Otherwise, the more important difference is their eigenvalues. The eigenvalues of both the beam equation and the wave equation are on the imaginary axis, but the eigenvalues of the wave equation are equidistant, and the eigenvalues of the beam equation grow further and further along the imaginary axis. These differences also lead to differences in the control of the two equations.

Remark 2.1 For clarity, notations $(*)_x = \frac{\partial(*)}{\partial x}$, $(*)_{xx} = \frac{\partial^2(*)_x}{\partial x^2}$, $(*)_{xxx} = \frac{\partial^3(*)}{\partial x^3}$, $(*)_{xxxx} = \frac{\partial^4(*)}{\partial x^4}$, $(*)_t = \frac{\partial(*)}{\partial t}$, $(*)_{tt} = \frac{\partial^2(*)}{\partial t^2}$ are used throughout this book.

2.2   Aerodynamic Forces

(1) The aerodynamic forces are distributed loads acting on the hose, which are the results of the skin friction and the pressure drag. These drag forces are the functions of the flow velocities relative to the hose [5]. The skin friction drag $f_t(x, t)$ in the tangential direction can be written as

$$f_t(x, t) = C_f \frac{1}{2} \rho_{air} v_t^2 \pi D_h \tag{2.6}$$

where C_f is the skin friction coefficient of the flexible hose, ρ_{air} is the local air density, v_t represents the tangential component of the relative flow velocity $v(t)$, D_h is the diameter of the flexible hose.

The pressure drag $f_n(x, t)$ in the normal direction can be expressed as

$$f_n(x, t) = C_d \frac{1}{2} \rho_{air} v_n^2 D_h \tag{2.7}$$

where C_d is the pressure drag coefficient of the flexible hose, v_n describes the normal component of the relative flow velocity $v(t)$.

(2) The external force of the drogue

$$f_{drog}(t) = \frac{1}{2}\rho_{air}v^2(t)C_{drog}\frac{\pi D_{drog}^2}{4} \tag{2.8}$$

where D_{drog} is the diameter of the drogue, and C_{drog} is the drag coefficient, which depends on the physical characteristics of the drogue.

2.3 The Hamilton Principle

Compared with the mechanical system described by lumped-parameter, the flexible mechanical system is essentially an infinitely dimensional distributed parameter system, which is a continuous function with respect to space and time. According to Hamilton principle, the dynamic model of the flexible system can be obtained from the kinetic energy, potential energy and virtual work of the flexible system by variation. Hamilton principle [1, 2] can be expressed as

$$\int_{t_1}^{t_2} (\delta E_k(t) - \delta E_p(t) + \delta W(t))dt = 0 \tag{2.9}$$

where $\delta(\cdot)$ denotes the variational operator, t_1 and t_2 are two time instants which satisfy $t_1 < t < t_2$, $E_k(t)$ and $E_p(t)$ represent the kinetic energy and potential energy of the system, respectively. $\delta W(t)$ is the virtual work done by non-conservative forces of the system, usually including internal tension, transverse load, linear structural damping and external disturbances.

There are several advantages of establishing the dynamic model of the flexible mechanical system through Hamilton principle. Firstly, this method is independent of the selection of coordinate system and can obtain the boundary conditions of the model at the same time; Secondly, the kinetic energy, potential energy and virtual work required by Hamilton principle can be directly used in the selection of Lyapunov function for the proof of stability of the closed-loop system.

2.4 Separation of Variables

Separation of Variables is the most commonly used method to solve constant coefficient PDEs. We introduce the application of Separation of Variables through a simple example. Consider the following diffusion equation which includes a reaction term

$$w_t(x, t) = w_{xx}(x, t) + \lambda w(x, t) \tag{2.10}$$

with boundary conditions

$$w(0, t) = 0 \tag{2.11}$$
$$w(1, t) = 0 \tag{2.12}$$

and initial condition $w(x, 0) = w_0(x)$. Then let us find the solution to this system. Assume that the solution of the system $w(x, t)$ can be written as a product of a function of space and a function of time,

$$w(x, t) = X(x)T(t) \tag{2.13}$$

Substituting (2.13) into (2.10), we can obtain

$$X(x)\dot{T}(t) = X''(x)T(t) + \lambda X(x)T(t) \tag{2.14}$$

Gathering the like terms on the opposite sides yields

$$\frac{\dot{T}(t)}{T(t)} = \frac{X''(x) + \lambda X(x)}{X(x)} \tag{2.15}$$

Since the function on the left depends only on time, and the function on the right depends only on the spatial variable, the equality can hold only if both functions are constant. Let us denote this constant by σ, we further get two ODEs

$$\dot{T} = \sigma T \tag{2.16}$$

with initial condition is $T(0) = T_0$, and

$$X'' + (\lambda - \sigma)X = 0 \tag{2.17}$$

with the boundary conditions $X(0) = X(1) = 0$ (following from the PDE boundary conditions). The solution to (2.16) is given by

$$T = T_0 e^{\sigma t} \tag{2.18}$$

The solution to (2.17) is

$$X(x) = A \sin(\sqrt{\lambda - \sigma} x) + B \cos(\sqrt{\lambda - \sigma} x) \tag{2.19}$$

where A and B are two constants that should be determined from the boundary conditions. We then have $B = 0$, $A \sin(\sqrt{\lambda - \sigma}) = 0$.

The last equality can hold only if $\sqrt{\lambda - \sigma} = \pi n, n = 0, 1, 2, \ldots$, so that

$$\sigma = \lambda - \pi^2 n^2, \ n = 0, 1, 2, \ldots \tag{2.20}$$

Substituting (2.18) and (2.19) into (2.13), we have

$$w_n(x, t) = T_0 A_n e^{(\lambda - \pi^2 n^2)} \sin(\pi n x), \ n = 0, 1, 2, \ldots \tag{2.21}$$

For linear PDEs, the sum of particular solutions is also a solution (the principle of superposition). Therefore the formal general solution is given by

$$w(x, t) = \sum_{n=1}^{\infty} C_n e^{(\lambda - \pi^2 n^2)} \sin(\pi n x) \tag{2.22}$$

where $C_n = A_n T_0$.

2.5 Useful Lemmas

2.5.1 Lyapunov's Direct Method

Lemma 2.1 *Suppose that a functional $V(w), t \geq 0, x \in D, \exists \alpha, \beta > 0$ make sure that $\beta \|w\|^2 \geq V(w) \geq \alpha \|w\|^2$ and $\dot{V} \leq 0$, then the function values $w(x, t) = 0$ is stable. If $V \to 0$ when $t \to \infty$, then $w(x, t) = 0$ is asymptotically stable. At the same time, if $\dot{V} \leq -\lambda \|w\|^2$ is also satisfied, then $w(x, t) = 0$ is exponentially stable. Where D is the domain of definition and $\| \cdot \|$ is the standard norm in $\mathbb{R}^n$.*

From the conditions of exponential stability, we have

$$\dot{V} \leq -\lambda \|w\|^2 \leq \frac{\lambda}{\beta} V = -\gamma V \tag{2.23}$$

Integrating of (2.23), we can obtain

$$V(t) \leq V(0) e^{-\gamma t} \tag{2.24}$$

We further obtain

$$\|w(x, t)\| \leq \sqrt{\frac{\beta |w(x, 0)|^2}{\alpha} e^{-\gamma t}} \tag{2.25}$$

And $w(x, t)$ decays to zero exponentially.

Lemma 2.2 *Suppose that a positive scalar functional $V(w)$, if*

$$\dot{V} \leq -\gamma V + \varepsilon \tag{2.26}$$

$\gamma, \varepsilon > 0$, *then* $\forall t \in [0, \infty)$, *the following inequality holds:*

$$V \leq V(0)e^{-\gamma t} + \frac{\varepsilon}{\gamma}(1 - e^{-\gamma t}) \tag{2.27}$$

2.5.2   Barbalat's Lemma

If the derivative of a system's Lyapunov function with respect to time is semidefinite, the asymptotically stable of the system can be obtained by Barbalat's Lemma.

Lemma 2.3 *Let $V(t) \in \mathbb{R}$ be a non-negative function of time on $[0, \infty]$. If (i) the time derivative of $V(t)$ satisfy $\dot{V}(t) \leq -f(t)$, where $f(t)$ is a non-negative function, and (ii) $f(t)$ is uniformly continuous (or if $\dot{f}(t) \in \mathcal{L}_\infty$), then*

$$\lim_{t \to \infty} f(t) = 0 \tag{2.28}$$

2.5.3   Other Useful Lemmas

Lemma 2.4 ([4]) *Let $\phi_1(x, t), \phi_2(x, t) \in \mathbb{R}, \forall(x, t) \in [0, L] \times [0, \infty)$, the following inequalities hold:*

$$\phi_1(x, t)\phi_2(x, t) \leq |\phi_1(x, t)\phi_2(x, t)| \leq \phi_1^2(x, t) + \phi_2^2(x, t) \tag{2.29}$$

$$\phi_1(x, t)\phi_2(x, t) \leq \frac{1}{\gamma}\phi_1^2(x, t) + \gamma\phi_2^2(x, t) \tag{2.30}$$

where γ is a positive constant.

Lemma 2.5 ([4]) *Let $\phi(x, t) \in \mathbb{R}, \forall(x, t) \in [0, L] \times [0, \infty)$, If $\phi(0, t) = 0, \forall t \in [0, \infty)$, the following inequality holds:*

$$\phi^2(x, t) \leq L \int_0^L [\phi_x(x, t)]^2 dx, \forall x \in [0, L] \tag{2.31}$$

Similarly, if $\phi_x(0, t) = 0, \forall t \in [0, \infty)$, then

$$[\phi_x(x, t)]^2 \leq L \int_0^L [\phi_{xx}(x, t)]^2 dx, \forall x \in [0, L] \tag{2.32}$$

Lemma 2.6 *Poincaré inequalities: For any $\phi(x, t)$ continuously differentiable on $[L_1, L_2]$, we have*

$$\int_{L_1}^{L_2} [\phi(x, t)]^2 dx \leq 2(L_2 - L_1)\phi^2(L_2, t) + 4(L_2 - L_1)^2 \int_{L_1}^{L_2} [\phi_x(x, t)]^2 dx$$

(2.33)

$$\int_{L_1}^{L_2} [\phi(x, t)]^2 dx \leq 2(L_2 - L_1)\phi^2(L_1, t) + 4(L_2 - L_1)^2 \int_{L_1}^{L_2} [\phi_x(x, t)]^2 dx$$

(2.34)

Proof Using integration by parts, we have

$$2 \int_{L_1}^{L_2} (x - L_1)\, \phi(x, t)\phi_x(x, t)dx = (x - L_1)\, \phi^2(x, t)|_{L_1}^{L_2} - \int_{L_1}^{L_2} [\phi(x, t)]^2 dx$$

$$= (L_2 - L_1)\phi^2(L_2, t) - \int_{L_1}^{L_2} [\phi(x, t)]^2 dx$$

(2.35)

We further have

$$\int_{L_1}^{L_2} [\phi(x, t)]^2 dx = (L_2 - L_1)\phi^2(L_2, t) - 2 \int_{L_1}^{L_2} (x - L_1)\, \phi(x, t)\phi_x(x, t)dx$$

$$\leq (L_2 - L_1)\phi^2(L_2, t) + \frac{1}{2} \int_{L_1}^{L_2} [\phi(x, t)]^2 dx + 2 \int_{L_1}^{L_2} (x - L_1)^2 [\phi_x(x, t)]^2 dx$$

$$\leq (L_2 - L_1)\phi^2(L_2, t) + \frac{1}{2} \int_{L_1}^{L_2} [\phi(x, t)]^2 dx + 2(L_2 - L_1)^2 \int_{L_1}^{L_2} [\phi_x(x, t)]^2 dx$$

(2.36)

Then we obtain

$$\int_{L_1}^{L_2} [\phi(x, t)]^2 dx \leq 2(L_2 - L_1)\phi^2(L_2, t) + 4(L_2 - L_1)^2 \int_{L_1}^{L_2} [\phi_x(x, t)]^2 dx$$

(2.37)

Inequality (2.34) is obtained in a similar fashion.

Remark 2.2 From Poincaré inequalities (2.33) and (2.34), we further have

$$\int_{L_1}^{L_2} [\phi(x, t)]^2 dx \leq 2(L_2 - L_1)\phi^2(L_2, t) + 8(L_2 - L_1)^3 \phi_x^2(L_2, t)$$

$$+ 16(L_2 - L_1)^4 \int_{L_1}^{L_2} [\phi_{xx}(x, t)]^2 dx \tag{2.38}$$

$$\int_{L_1}^{L_2} [\phi(x, t)]^2 dx \leq 2(L_2 - L_1)\phi^2(L_1, t) + 8(L_2 - L_1)^3 \phi_x^2(L_1, t)$$

$$+ 16(L_2 - L_1)^4 \int_{L_1}^{L_2} [\phi_{xx}(x, t)]^2 dx \tag{2.39}$$

Lemma 2.7 ([3]) *The following inequality holds for any $\varepsilon > 0$ and for any $\chi \in R$:*

$$0 \leq |\chi| - \chi \tanh\left(\frac{\chi}{\varepsilon}\right) \leq \mu\varepsilon, \mu = 0.2785 \tag{4.12}$$

Lemma 2.8 ([6]) *Let $V(\cdot)$ and $\chi(\cdot)$ be smooth functions defined on $[0, t_f)$ with $V(t) \geq 0$, $\forall t \in [0, t_f)$ and $N(\chi)$ be an even smooth Nussbaum function. The following inequality holds:*

$$V \leq V(0) e^{-Ct} c_0 + \frac{M}{C}\left(1 - e^{-Ct}\right) + \frac{e^{-Ct}}{\gamma_\chi} \int_0^t (\zeta N(\varsigma)\dot{\chi} - \dot{\chi}) e^{-Ct} d\tau \tag{2.40}$$

where $C > 0$, $M > 0$, $\gamma_\chi > 0$, $\zeta = \frac{\partial g(v)}{\partial v} = \frac{4}{\left(e^{v/u_M} + e^{-v/u_M}\right)^2} > 0$, then $V(\cdot)$ and $\chi(\cdot)$ are bounded on $[0, t_f)$.

References

1. Goldstein H (1951) Classical mechanics. Addison-Wesley, Massachusetts, USA
2. Meirovitch L (1967) Analytical methods in vibrations. The Macmillan Company, New York, NY
3. Polycarpou MM, Ioannou PA (1996) A robust adaptive nonlinear control design. Automatica 32(3):423–427
4. Rahn CD (2001) Mechatronic control of distributed noise and vibration. Springer, Berlin
5. Ro K, Kuk T, Kamman J (2010) Active control of aerial refueling hose-drogue systems. In AIAA Guidance, Navigation, and Control Conference. pp 8400
6. Wen C, Zhou J, Liu Z, Hongye S (2011) Robust adaptive control of uncertain nonlinear systems in the presence of input saturation and external disturbance. IEEE Trans Autom Control 56(7):1672–1678

Chapter 3
PDE Modeling and Basic Vibration Control for Flexible Satellite

3.1 Introduction

Satellites have gained much attention in the past decades for the purpose of remote sensing, communication and so on [8–10]. Considering a complex environment and space mission constraints, flexible satellite characterized by light weight structure, limitation of mass, low energy consumption and reduced launch cost, are widely used. Flexible satellites usually consists of a central rigid hub and long flexible appendages such as solar panels. However, due to the flexible property of flexible appendages, the unwanted deflection of the flexible modes and other external disturbances have a significant influence on the dynamics and control performance of the flexible satellite. Therefore, high-precision attitude control for flexible satellites is a difficult problem and an important research topic.

Recently, a number of methods have been developed for the control of a flexible satellite, including adaptive fuzzy sliding mode control [3], H-infinity control [1], variable structure attitude control [5, 6] and so on. However, the above mentioned papers only consider one flexible panel and the models are based on the ordinary differential equations (ODEs). As mentioned in the previous papers, spillover problem due to truncation of the model can lead to an unstable system, which should be avoided. There are significant research efforts for flexible structures where the control design is based on the original distributed parameter systems [2, 4, 11]. However, the model of flexible satellite with two solar panels (two Euler–Bernoulli beams) is different from the models in [2, 4, 11] that consists of a single Euler–Bernoulli beam, and the previous control methods cannot be applied to the problem in this chapter.

In this chapter, we aim to deal with the active vibration suppression problem for a flexible satellite. The configuration of the flexible satellite with two solar panels is shown in Fig. 3.1. The effect of rotation angle for the flexible satellite is not considered. Two Euler–Bernoulli beams connected to a center body are used to model the dynamics of flexible satellite. The left and the right panels are modeled as two beams, and the center body of the satellite is modeled as a lumped mass in the

© Tsinghua University Press 2020
Z. Liu and J. Liu, *PDE Modeling and Boundary Control for Flexible Mechanical System*, Springer Tracts in Mechanical Engineering,
https://doi.org/10.1007/978-981-15-2596-4_3

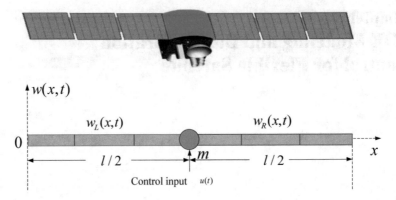

Fig. 3.1 Satellite with two symmetrical flexible solar panels

center of two panels. The flexible satellite systems can be actually regarded as a free-free beam with a point load in the center. The structure dynamics of flexible satellite belongs to the distributed parameter systems described by hybrid PDEs-ODEs, shown in Sect. 3.2. The control and control-related issues are presented through theoretical analysis and simulations. A single-point control input is proposed on the basis of the original distributed parameter system to control the deformation of both flexible panels. With the proposed control, the closed-loop system is exponentially stable via the Lyapunov's direct method. The control performance of the system is guaranteed by suitably choosing the control parameters.

The rest of this chapter is as follows. In Sect. 3.2, the dynamic model of flexible satellite is given for the subsequent development. Based on the Lyapunov stability theory, boundary control schemes are proposed to control the deformation of panels in Sect. 3.3, where it is shown that the exponential stability of the closed-loop system can be achieved by the proposed control. Simulations are carried out to illustrate the performance of the proposed control in Sect. 3.4.

3.2 Problem Formulation

The satellite dynamic model is composed of a center body with two identical flexible panels. Hamilton's principle is used to derive the equations of motion for the satellite, starting from the expression of the kinetic and potential energy of the system. Let $w(x, t)$ denote the transverse displacements of the panels from their initial equilibrium position at position x and time t, respectively, $w(l/2, t)$ denotes the transverse displacements of the lumped mass, ρ is the density of the beam material, A is the cross-sectional area of the beam, E is Young's modulus, I is the area moment of inertia of the beam, γ_1 is the coefficient of viscous damping, m is the mass of the center body, and $u(t)$ is a single-point controller at the center body. The actuator is located at the center body to regulate the vibrations of two flexible panels.

The kinetic energy of the beam $E_k(t)$ can be represented as

$$E_k(t) = \frac{1}{2}\rho A \int_0^{l/2} \left(\frac{\partial w_L(x,t)}{\partial t}\right)^2 dx + \frac{1}{2}m \left(\frac{\partial w(x,t)}{\partial t}\right)^2 \bigg|_{x=l/2}$$
$$+ \frac{1}{2}\rho A \int_{l/2}^l \left(\frac{\partial w_R(x,t)}{\partial t}\right)^2 dx \tag{3.1}$$

where x and t represent the independent spatial and time variables, respectively. The potential energy $E_p(t)$ due to the bending can be obtained from

$$E_p(t) = \frac{1}{2}EI \int_0^{l/2} \left(\frac{\partial^2 w_L(x,t)}{\partial x^2}\right)^2 dx + \frac{1}{2}EI \int_{l/2}^l \left(\frac{\partial^2 w_R(x,t)}{\partial x^2}\right)^2 dx \tag{3.2}$$

The virtual work done by damping on the system is represented by

$$\delta W_d = -\int_0^{l/2} \gamma_1 \frac{\partial w_L(x,t)}{\partial t} \delta w_L(x,t) dx - \int_{l/2}^l \gamma_1 \frac{\partial w_R(x,t)}{\partial t} \delta w_R(x,t) dx \tag{3.3}$$

The virtual work done by the axial control force $u(t)$ that produces a transverse force for vibration suppression can be written as

$$\delta W_u(t) = u(t)\delta w(l/2, t) \tag{3.4}$$

Then, we have the total virtual work done on the system as

$$\delta W(t) = \delta W_d(t) + \delta W_u(t) \tag{3.5}$$

The variations of (3.1) and (3.2) are obtained as

$$\delta E_k(t) = \rho A \int_0^{l/2} \frac{\partial w_L(x,t)}{\partial t} \delta \frac{\partial w_L(x,t)}{\partial t} dx + m \frac{\partial w(x,t)}{\partial t} \delta \frac{\partial w(x,t)}{\partial t} \bigg|_{x=l/2}$$
$$+ \rho A \int_{l/2}^l \frac{\partial w_R(x,t)}{\partial t} \delta \frac{\partial w_R(x,t)}{\partial t} dx \tag{3.6}$$

$$\delta E_p(t) = EI \int_0^{l/2} \frac{\partial^2 w_L(x,t)}{\partial x^2} \delta \frac{\partial^2 w_L(x,t)}{\partial x^2} dx$$
$$+ EI \int_{l/2}^l \frac{\partial^2 w_R(x,t)}{\partial x^2} \delta \frac{\partial^2 w_R(x,t)}{\partial x^2} dx \tag{3.7}$$

and we further obtain

$$
\int_{t_1}^{t_2} \delta E_k(t)\,dt = -\rho A \int_{t_1}^{t_2} \int_0^{l/2} \frac{\partial^2 w_L(x,t)}{\partial t^2} \delta w_L(x,t)\,dx\,dt
$$

$$
- m \int_{t_1}^{t_2} \frac{\partial^2 w(x,t)}{\partial t^2} \delta w(x,t) \bigg|_{x=l/2} dt
$$

$$
- \rho A \int_{t_1}^{t_2} \int_{l/2}^{l} \frac{\partial^2 w_R(x,t)}{\partial t^2} \delta w_R(x,t)\,dx\,dt \qquad (3.8)
$$

$$
\int_{t_1}^{t_2} \delta E_p(t)\,dt = EI \int_{t_1}^{t_2} \int_0^{l/2} \frac{\partial^4 w_L(x,t)}{\partial x^4} \delta w_L(x,t)\,dx\,dt
$$

$$
+ EI \int_{t_1}^{t_2} \int_{l/2}^{l} \frac{\partial^4 w_R(x,t)}{\partial x^4} \delta w_R(x,t)\,dx\,dt
$$

$$
+ EI \int_{t_1}^{t_2} \left[\frac{\partial^2 w_L(x,t)}{\partial x^2} \delta \frac{\partial w_L(x,t)}{\partial x} - \frac{\partial^3 w_L(x,t)}{\partial x^3} \delta w_L(x,t) \right] \bigg|_0^{l/2} dt
$$

$$
+ EI \int_{t_1}^{t_2} \left[\frac{\partial^2 w_R(x,t)}{\partial x^2} \delta \frac{\partial w_R(x,t)}{\partial x} - \frac{\partial^3 w_R(x,t)}{\partial x^3} \delta w_R(x,t) \right] \bigg|_{l/2}^{l} dt
$$

$$
(3.9)
$$

Applying Hamilton's principle,

$$
\int_{t_1}^{t_2} \delta \left[E_k(t) - E_p(t) + W(t) \right] dt = 0
$$

we obtain the following structure dynamics of the spacecraft with the governing equations as

$$
\rho A w_{Ltt}(x,t) + EI w_{Lxxxx}(x,t) + \gamma_1 w_{Lt}(x,t) = 0 \qquad (3.10)
$$

$\forall x \in [0, l/2], t \in [0, \infty)$, and

$$
\rho A w_{Rtt}(x,t) + EI w_{Rxxxx}(x,t) + \gamma_1 w_{Rt}(x,t) = 0 \qquad (3.11)
$$

$\forall x \in [l/2, l], t \in [0, \infty)$, and boundary conditions as

$$
w_{Lx}(l/2, t) = w_{Rx}(l/2, t) = 0 \qquad (3.12)
$$

$$
w_{Lxx}(0, t) = w_{Rxx}(l, t) = 0 \qquad (3.13)
$$

$$
w_{Lxxx}(0, t) = w_{Rxxx}(l, t) = 0 \qquad (3.14)
$$

$$w_L(l/2, t) = w_R(l/2, t) = w(l/2, t) \tag{3.15}$$

$$mw_{tt}(l/2, t) = EIw_{Lxxx}(l/2, t) - EIw_{Rxxx}(l/2, t) + u(t) \tag{3.16}$$

$t \in [0, \infty)$.

Remark 3.1 Boundary condition (3.15) is a motion equation of the center body in the satellite system. $w_{tt}(l/2, t)$ denotes acceleration of the center body, $EIw_{Lxxx}(l/2, t)$ describes the shear force from the left panel, $EIw_{Rxxx}(l/2, t)$ describes the shear force from the right panel, and $u(t)$ is the control force from the actuator.

3.3 Control Design

The control objective is to propose an active control law to regulate the deformation of the two flexible panels. A single-point control force $u(t)$ is applied on the center body of the satellite. Consider the Lyapunov candidate function as

$$V(t) = V_1(t) + V_2(t) + \Delta(t) \tag{3.17}$$

where $V_1(t)$, $V_2(t)$ and $\Delta(t)$ are defined as

$$
\begin{aligned}
V_1(t) = {}& \frac{\beta}{2}\rho A \int_0^{l/2} [w_{Lt}(x, t)]^2 dx + \frac{\beta}{2}EI \int_0^{l/2} [w_{Lxx}(x, t)]^2 dx \\
& + \frac{\alpha}{2}\gamma_1 \int_0^{l/2} [w_L(x, t)]^2 dx + \frac{\beta}{2}\rho A \int_{l/2}^l [w_{Rt}(x, t)]^2 dx \\
& + \frac{\beta}{2}EI \int_{l/2}^l [w_{Rxx}(x, t)]^2 dx + \frac{\alpha}{2}\gamma_1 \int_{l/2}^l [w_R(x, t)]^2 dx
\end{aligned} \tag{3.18}
$$

$$V_2(t) = \frac{\beta}{2}mS^2(t) + \frac{\beta k_p}{2}[w(l/2, t)]^2 \tag{3.19}$$

$$\Delta(t) = \alpha\rho A \int_0^{l/2} w_{Lt}(x, t)w_L(x, t)dx + \alpha\rho A \int_{l/2}^l w_{Rt}(x, t)w_R(x, t)dx \tag{3.20}$$

where α and β are positive weighting constants, k_p is the control gain, and

$$S(t) = \frac{\alpha}{\beta}w(l/2, t) + w_t(l/2, t) \tag{3.21}$$

$V_1(t)$ is bounded as

$$V_1(t) \geq \theta_1 \left[\int_0^{l/2} \left([w_{Lt}(x,t)]^2 + [w_L(x,t)]^2 \right) dx + \int_{l/2}^l \left([w_{Rt}(x,t)]^2 + [w_R(x,t)]^2 \right) dx \right] \tag{3.22}$$

where $\theta_1 = \min\left(\frac{\beta \rho A}{2}, \frac{\alpha \gamma_1}{2} \right) > 0$.

From the definition of $\Delta(t)$, we know $\Delta(t)$ is bounded as

$$|\Delta(t)| \leq \alpha \rho A \left[\int_0^{l/2} \left([w_{Lt}(x,t)]^2 + [w_L(x,t)]^2 \right) dx + \int_{l/2}^l \left([w_{Rt}(x,t)]^2 + [w_R(x,t)]^2 \right) dx \right]$$

$$\leq \theta_2 V_1(t) \tag{3.23}$$

where $\theta_2 = \frac{\alpha \rho A}{\theta_1}$. Considering $\theta_1 > \alpha \rho A$, we can have

$$0 \leq \theta_4 V_1(t) \leq V_1(t) + \Delta(t) \leq \theta_3 V_1(t) \tag{3.24}$$

where $\theta_3 = 1 + \theta_2 > 1$ and $0 < \theta_4 = 1 - \theta_2 < 1$. Then considering the Lyapunov candidate function (3.17), we have

$$0 \leq \lambda_2 \left[V_1(t) + V_2(t) \right] \leq V(t) \leq \lambda_1 \left[V_1(t) + V_2(t) \right] \tag{3.25}$$

where $\lambda_1 = \max(\theta_3, 1) = \theta_3$, $\lambda_2 = \min(\theta_4, 1) = \theta_4$.

Differentiating $V(t)$ leads to

$$\dot{V}(t) = \dot{V}_1(t) + \dot{V}_2(t) + \dot{\Delta}(t) \tag{3.26}$$

where $\dot{V}_1(t)$ is given as

$$\dot{V}_1(t) = \beta \rho A \int_0^{l/2} w_{Lt}(x,t) w_{Ltt}(x,t) dx + \beta EI \int_0^{l/2} w_{Lxx}(x,t) w_{Lxxt}(x,t) dx$$

$$+ \alpha \gamma_1 \int_0^{l/2} w_L(x,t) w_{Lt}(x,t) dx + \beta \rho A \int_{l/2}^l w_{Rt}(x,t) w_{Rtt}(x,t) dx$$

$$+ \beta EI \int_{l/2}^l w_{Rxx}(x,t) w_{Rxxt}(x,t) dx + \alpha \gamma_1 \int_{l/2}^l w_R(x,t) w_{Rt}(x,t) dx \tag{3.27}$$

Substituting the governing equations (3.10) and (3.11), we obtain

$$\dot{V}_1(t) = A_1(t) + A_2(t) + A_3(t) + A_4(t) \tag{3.28}$$

where

$$A_1(t) = -\beta EI \int_0^{l/2} w_{Lt}(x,t) w_{Lxxxx}(x,t) dx + \beta EI \int_0^{l/2} w_{Lxx}(x,t) w_{Lxxt}(x,t) dx \tag{3.29}$$

$$A_2(t) = -\beta E I \int_{l/2}^{l} w_{Rt}(x,t)w_{Rxxxx}(x,t)dx + \beta E I \int_{l/2}^{l} w_{Rxx}(x,t)w_{Rxxt}(x,t)dx \tag{3.30}$$

$$A_3(t) = -\beta\gamma_1 \int_{0}^{l/2} [w_t(x,t)]^2 dx - \beta\gamma_1 \int_{l/2}^{l} [w_t(x,t)]^2 dx \tag{3.31}$$

$$A_4(t) = \alpha\gamma_1 \int_{0}^{l/2} w_L(x,t)w_{Lt}(x,t)dx + \alpha\gamma_1 \int_{l/2}^{l} w_R(x,t)w_{Rt}(x,t)dx \tag{3.32}$$

Using integration by parts and boundary conditions (3.12) and (3.13), we have

$$
\begin{aligned}
A_1(t) = & -\beta E I w_{Lt}(l/2,t)w_{Lxxx}(l/2,t) + \beta E I w_{Lt}(0,t)w_{Lxxx}(0,t) \\
& + \beta E I w_{Lxt}(l/2,t)w_{Lxx}(l/2,t) - \beta E I w_{Lxt}(0,t)w_{Lxx}(0,t) \\
= & -\beta E I w_{Lt}(l/2,t)w_{Lxxx}(l/2,t)
\end{aligned}
\tag{3.33}
$$

$$
\begin{aligned}
A_2(t) = & -\beta E I w_{Rt}(l/2,t)w_{Rxxx}(l/2,t) + \beta E I w_{Rt}(0,t)w_{Rxxx}(0,t) \\
& + \beta E I w_{Rxt}(l/2,t)w_{Rxx}(l/2,t) - \beta E I w_{Rxt}(0,t)w_{Rxx}(0,t) \\
= & -\beta E I w_{Rt}(l/2,t)w_{Rxxx}(l/2,t)
\end{aligned}
\tag{3.34}
$$

Combining $A_1(t) - A_4(t)$ and applying boundary conditions (3.12), (3.14) and (3.15), we obtain the derivative of $V_1(t)$ as

$$
\begin{aligned}
\dot{V}_1(t) \leq & -\beta\gamma_1 \int_{0}^{l/2} [w_{Lt}(x,t)]^2 dx - \beta\gamma_1 \int_{l/2}^{l} [w_{Rt}(x,t)]^2 dx \\
& + \alpha\gamma_1 \int_{0}^{l/2} w_L(x,t)w_{Lt}(x,t)dx + \alpha\gamma_1 \int_{l/2}^{l} w_R(x,t)w_{Rt}(x,t)dx \\
& + \beta w_t(l/2,t) [E I w_{Lxxx}(l/2,t) - E I w_{Rxxx}(l/2,t)]
\end{aligned}
\tag{3.35}
$$

The derivative of $V_2(t)$ is given as

$$\dot{V}_2(t) = \beta m S(t)\dot{S}(t) + \beta k_p w(l/2,t)w_t(l/2,t) \tag{3.36}$$

Using boundary condition (3.16), we have

$$
\begin{aligned}
\dot{V}_2(t) = & \beta S(t)\left[u(t) + E I w_{Lxxx}(l/2,t) - E I w_{Rxxx}(l/2,t) + \frac{\alpha}{\beta}m w_t(l/2,t)\right] \\
& + \beta k_p w(l/2,t)w_t(l/2,t)
\end{aligned}
\tag{3.37}
$$

Design the proposed control scheme as

$$u(t) = -kS(t) - \frac{\alpha}{\beta}mw_t\,(l/2, t) - k_p w\,(l/2, t) \tag{3.38}$$

where $k > 0$ and $k_p > 0$ are the control gains. Using the above control, we have

$$
\begin{aligned}
\dot{V}_2(t) = {} & [\alpha w\,(l/2, t) + \beta w_t\,(l/2, t)]\,[EIw_{Lxxx}(l/2, t) - EIw_{Rxxx}(l/2, t)] \\
& - k\beta S^2(t) - \alpha k_p[w(l/2, t)]^2
\end{aligned}
\tag{3.39}
$$

The derivative of $\Delta(t)$ is given as

$$
\begin{aligned}
\dot{\Delta}(t) = {} & \alpha\rho A \int_0^{l/2} w_{Ltt}(x, t)w_L(x, t)dx + \alpha\rho A \int_0^{l/2} [w_{Lt}(x, t)]^2 dx \\
& + \alpha\rho A \int_{l/2}^{l} w_{Rtt}(x, t)w_R(x, t)dx + \alpha\rho A \int_{l/2}^{l} [w_{Rt}(x, t)]^2 dx
\end{aligned}
\tag{3.40}
$$

Substituting the governing equations (3.10) and (3.11), we obtain

$$\dot{\Delta}(t) = B_1(t) + B_2(t) + B_3(t) \tag{3.41}$$

where

$$B_1(t) = -\alpha EI \int_0^{l/2} w_L(x, t)w_{Lxxxx}(x, t)dx - \alpha EI \int_{l/2}^{l} w_R(x, t)w_{Rxxxx}(x, t)dx \tag{3.42}$$

$$B_2(t) = \alpha\rho A \int_0^{l/2} [w_{Lt}(x, t)]^2 dx + \alpha\rho A \int_{l/2}^{l} [w_{Rt}(x, t)]^2 dx \tag{3.43}$$

$$B_3(t) = -\alpha\gamma_1 \int_0^{l/2} w_L(x, t)w_{Lt}(x, t)dx - \alpha\gamma_1 \int_{l/2}^{l} w_R(x, t)w_{Rt}(x, t)dx \tag{3.44}$$

Using integration by parts and boundary conditions (3.12), (3.14) and (3.15), we obtain

$$
\begin{aligned}
B_1(t) = {} & -\alpha w(l/2, t)\,[EIw_{Lxxx}(l/2, t) - EIw_{Rxxx}(l/2, t)] \\
& - \alpha EI \int_0^{l/2} [w_{Lxx}(x, t)]^2 dx - \alpha EI \int_{l/2}^{l} [w_{Rxx}(x, t)]^2 dx
\end{aligned}
\tag{3.45}
$$

Combining $B_1(t) - B_3(t)$, we obtain

$$\dot{\Delta}(t) \le - \alpha w(l/2, t) [EI w_{Lxxx}(l/2, t) - EI w_{Rxxx}(l/2, t)] - \alpha EI \int_0^{l/2} [w_{Lxx}(x, t)]^2 dx$$
$$- \alpha EI \int_{l/2}^l [w_{Rxx}(x, t)]^2 dx + \alpha \rho A \int_0^{l/2} [w_{Lt}(x, t)]^2 dx + \alpha \rho A \int_{l/2}^l [w_{Rt}(x, t)]^2 dx$$
$$- \alpha \gamma_1 \int_0^{l/2} w_L(x, t) w_{Lt}(x, t) dx - \alpha \gamma_1 \int_{l/2}^l w_R(x, t) w_{Rt}(x, t) dx \tag{3.46}$$

Therefore, the derivative of the Lyapunov candidate function is given as

$$\dot{V}(t) \le - (\gamma_1 \beta - \alpha \rho A) \int_0^{l/2} [w_{Lt}(x, t)]^2 dx - (\gamma_1 \beta - \alpha \rho A) \int_{l/2}^l [w_{Rt}(x, t)]^2 dx$$
$$- \alpha EI \int_0^{l/2} [w_{Lxx}(x, t)]^2 dx - \alpha EI \int_{l/2}^l [w_{Rxx}(x, t)]^2 dx$$
$$- k\beta S^2(t) - \alpha k_p [w(l/2, t)]^2 \tag{3.47}$$

From inequalities (2.38) and (2.39) and applying boundary conditions (3.12) and (3.15), we have

$$\int_0^{l/2} [w_L(x, t)]^2 dx \le l[w_L(l/2, t)]^2 + l^3 [w_{Lt}(l/2, t)]^2 + l^4 \int_0^{l/2} [w_{Lxx}(x, t)]^2 dx$$
$$\le l[w(l/2, t)]^2 + l^4 \int_0^{l/2} [w_{Lxx}(x, t)]^2 dx \tag{3.48}$$

$$\int_{l/2}^l [w_R(x, t)]^2 dx \le l[w_R(l/2, t)]^2 + l^3 [w_{Rt}(l/2, t)]^2 + l^4 \int_{l/2}^l [w_{Rxx}(x, t)]^2 dx$$
$$\le l[w(l/2, t)]^2 + l^4 \int_{l/2}^l [w_{Rxx}(x, t)]^2 dx \tag{3.49}$$

Then we can obtain the following inequalities

$$-\eta_1 l[w(l/2, t)]^2 \le -\eta_1 \int_0^{l/2} [w_L(x, t)]^2 dx + \eta_1 l^4 \int_0^{l/2} [w_{Lxx}(x, t)]^2 dx \tag{3.50}$$

$$-\eta_2 l[w(l/2, t)]^2 \le -\eta_2 \int_{l/2}^l [w_R(x, t)]^2 dx + \eta_2 l^4 \int_{l/2}^l [w_{Rxx}(x, t)]^2 dx \tag{3.51}$$

where η_1 and η_2 are positive constants. Then we obtain

$$\dot{V}(t) \le -\left(\gamma_1\beta - \alpha\rho A\right) \int_0^{l/2} [w_{Lt}(x, t)]^2 dx - \left(\alpha EI - \eta_1 l^4\right) \int_0^{l/2} [w_{Lxx}(x, t)]^2 dx$$

$$- \eta_1 \int_0^{l/2} [w_L(x, t)]^2 dx - \left(\gamma_1\beta - \alpha\rho A\right) \int_{l/2}^l [w_{Rt}(x, t)]^2 dx$$

$$- \left(\alpha EI - \eta_2 l^4\right) \int_{l/2}^l [w_{Rxx}(x, t)]^2 dx - \eta_2 \int_{l/2}^l [w_R(x, t)]^2 dx - k\beta S^2(t)$$

$$- \left(\alpha k_p - \eta_1 l - \eta_2 l\right) [w(l/2, t)]^2 \tag{3.52}$$

We further have

$$\dot{V}(t) \le -\lambda_3 [V_1(t) + V_2(t)] \tag{3.53}$$

where

$$\lambda_3 = \min\left(\frac{2\gamma_1\beta - 2\alpha\rho A}{\beta\rho A}, \frac{2\alpha EI - 2\eta_1 l^4}{\beta EI}, \frac{2\alpha EI - 2\eta_2 l^4}{\beta EI}, \frac{2\eta_1}{\alpha\gamma_1}, \frac{2\eta_2}{\alpha\gamma_1}, \frac{2k}{m},\right.$$
$$\left.\frac{2\alpha k_p - 2\eta_1 l - 2\eta_2 l}{\beta k_p}\right) > 0 \tag{3.54}$$

Combining (3.25) and (3.53), we have

$$\dot{V}(t) \le -\lambda V(t) \tag{3.55}$$

where $\lambda = \lambda_3/\lambda_1$. We then can obtain the following theorem.

Theorem 3.1 *For the dynamical system described by governing equations (3.10) and (3.11) and boundary conditions (3.12)–(3.15), under the proposed boundary control (3.38), if the initial conditions are bounded, then the closed-loop system is exponentially stable.*

Remark 3.2 How to construct the control law from the Lyapunov function is the main issue of this chapter. First, we can obtain the energy term $V_1(t)$ from the analysis in Sect. 3.2. Because the Lyapunov candidate function $V(t)$ needs to be positive definite and $\dot{V}(t)$ satisfies $\dot{V}(t) \le -\lambda V(t)$, we design an auxiliary term $V_2(t)$ and a crossing term $\Delta(t)$. We can then design the control law $u(t)$ for vibration suppression and substitute it in $\dot{V}_2(t)$. In turn, substitute the governing equations and boundary conditions in $\dot{V}(t)$ and see what term should be added in the Lyapunov function $V(t)$ and control law $u(t)$ in order to satisfy $\dot{V}(t) \le -\lambda V(t)$. After continuous revision and calculation of the Lyapunov function and control law, we can obtain the appropriate $V(t)$ and $u(t)$ to achieve the control objective.

Table 3.1 Parameters of the flexible satellite system

Parameter	Description	Value
$l/2$	Length of the panel	$10\,\mathrm{m}$
m	Mass of the centrebody	$20\,\mathrm{kg}$
ρ	Density of the material	$2.7 \times 10^3\,\mathrm{kg/m^3}$
A	Cross-sectional area of the panel	$0.12\,\mathrm{m^2}$
E	Young's modulus	$6.894 \times 10^{10}\,\mathrm{N/m^2}$
I	Area moment of inertia of the panel	$1.734 \times 10^{-7}\,\mathrm{m^4}$
γ_1	Viscous damping	$0.005\,\mathrm{kg(ms)}$

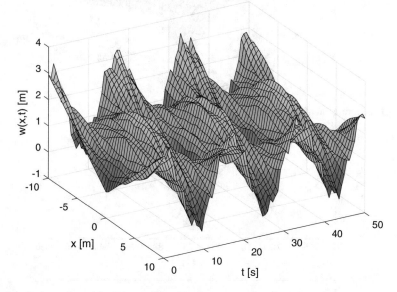

Fig. 3.2 Deformation of flexible satellite without control

3.4 Simulation

We select the finite difference method [7] to simulate the system performance with the proposed boundary control. By choosing the proper temporal and spatial step size to approximate the solution of the PDE model, the performance of the proposed control is well demonstrated via the finite difference method.

The initial conditions of the flexible satellite are given as $w_L(x, 0) = -0.3x$, $w_R(x, 0) = 0.3x$, $w_{Lt}(x, 0) = w_{Rt}(x, 0) = 0$. Detailed parameters of flexible satellite system are given in Table 3.1.

Displacement of the flexible satellite with control

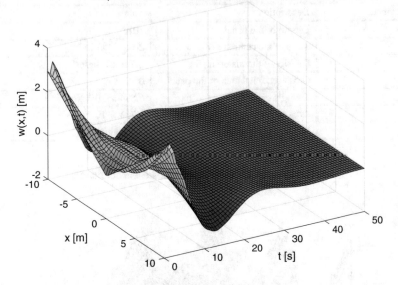

Fig. 3.3 Deformation of flexible satellite with proposed control

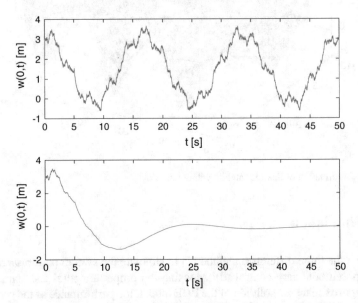

Fig. 3.4 Boundary displacement $w(0, t)$ of flexible satellite: without control and with control

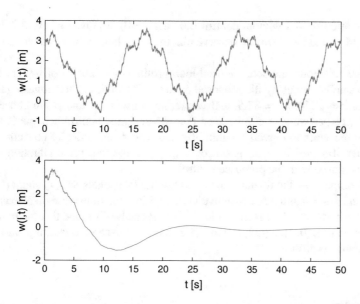

Fig. 3.5 Boundary displacement $w(l, t)$ of flexible satellite: without control and with control

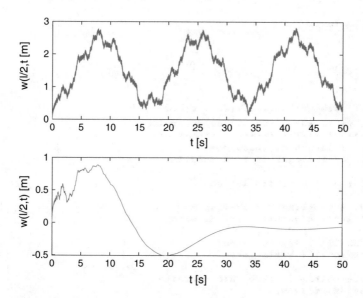

Fig. 3.6 Displacement of center body $w(l/2, t)$ of flexible satellite: without control and with control

Simulation results without control (i.e., $u(t) = 0$) are given in Fig. 3.2. For the results presented here, we can observe that there are large vibrations along the two panels.

Figure 3.3 shows the actual closed-loop profiles of evolution of $w(x, t)$ for the flexible satellite by using the proposed control. The values of the control gains are given as $k = 1250$, $k_p = 37.5$, and weighting constants α and β are chosen as 1 and 100, respectively. It can be seen that the designed control scheme is able to regulate the vibration greatly within 30 s and $w(x, t)$ numerically converge to the zero after 40 s, which means good convergence of the transverse vibration $w(x, t)$ can be achieved with the proposed control.

For comparison, the tip deformation of the flexible panels $w(0, t)$ and $w(l, t)$ are shown in Figs. 3.4 and 3.5, respectively. In addition, the transverse displacement of the center body $w(l/2, t)$ is shown in Fig. 3.6. It can be observed that the transverse displacements $w(0, t)$, $w(l/2, t)$ and $w(l, t)$ converge to zero, illustrating that control performance is ensured.

Appendix

Simulation Program:

```
1   clc;
2   close all;
3   clear all;
4
5   %****************************************
6   %     Flexible satellite withou control
7   %****************************************
8   nx=50;  % length of space domain
9   nt=8*10^4; % length of time domain
10
11  tmax=50; % time of simulation
12  L=10;
13  Ttr=80; % sampling for drawing
14  Ttr1=1*10^3; % sampling for drawing
15
16  dx=L/(nx-1); % spacing step
17  dt=tmax/(nt-1); % time step
18
19  % create matrix to save data contains
20  w=zeros(nx,nt);v=w;
21  u=zeros(nt,1);
22  w_3D=zeros(Ttr,nx);v_3D=w_3D;
23  t_3D=zeros(Ttr,2*nx-1);
24  revise_free=zeros(Ttr,nx-1);
25
26
27  % parameters
28  m=20;
29  A=0.12;
30  % E=5*10^8;
```

```
31   E=6.894*10^10;
32   % rho=5.24;
33   rho=2.7*10^3;
34   I=1.734*10^-7;
35   gamma1-0.005;
36   rhoA=rho*A;
37   EI=E*I;
38
39   % initial condition
40   for i=1:nx
41   w(i,1)=( nx-i )*dx*0.3;
42   end
43
44   w(:,2)=w(:,1);
45   w(:,3)=w(:,1);
46
47   for i=1:nx
48   v(i,1)=( i-1)*dx*0.3;
49   end
50   v(:,2)=v(:,1);
51   v(:,3)=v(:,1);
52
53   wl_free=zeros(nt,1);vl_free=wl_free;
54   wl_free(1)=w(1,1);wl_free(2)=wl_free(1);
55   vl_free(1)=v(nx,1);vl_free(2)=v(1);
56   w_L_free=zeros(nt,1);
57   w_L_free(1)=w(nx,1);
58   w_L_free(2)=w(nx,1);
59
60   %main cycle
61
62   for j=3:nt-1
63
64   for i=3:nx-2
65
66   wxxxx=( w(i+2,j)-4*w(i+1,j)+6*w(i,j)-4*w(i-1,j)+w(i-2,j) )/dx^4;
67   dw=( w(i,j)-w(i,j-1) )/dt;
68   vxxxx=( v(i+2,j)-4*v(i+1,j)+6*v(i,j)-4*v(i-1,j)+v(i-2,j) )/dx^4;
69   dv=( v(i,j)-v(i,j-1) )/dt;
70
71   w(i,j+1)=2*w(i,j)-w(i,j-1)+( -EI*wxxxx-gamma1*dw )*dt^2/rhoA;
72   v(i,j+1)=2*v(i,j)-v(i,j-1)+( -EI*vxxxx-gamma1*dv )*dt^2/rhoA;
73   end
74
75   w(1,j+1)=3*w(3,j+1)-2*w(4,j+1);
76   w(2,j+1)=2*w(3,j+1)-w(4,j+1);
77   v(nx,j+1)=3*v(nx-2,j+1)-2*v(nx-3,j+1);
78   v(nx-1,j+1)=2*v(nx-2,j+1)-v(nx-3,j+1);
79
80
81   wxxxl=( w(nx,j)-3*w(nx-1,j)+3*w(nx-2,j)-w(nx-3,j))/dx^3;
82
83   vxxxl=( v(4,j)-3*v(3,j)+3*v(2,j)-v(1,j) )/dx^3;
84   S(j)=wxxxl-vxxxl;
85   Q(j)=wxxxl;
86   R(j)=vxxxl;
87   w(nx,j+1)=2*w(nx,j)-w(nx,j-1)+( EI*wxxxl-EI*vxxxl+u(j-1) )*dt^2/m;
88   v(1,j+1)=w(nx,j+1);
89
```

```
90   w(nx-1,j+1)=w(nx,j+1);
91   v(2,j+1)=v(1,j+1);
92
93   w1_free(j)=w(1,j+1);
94   v1_free(j)=v(nx,j+1);
95   w_L_free(j)=w(nx,j+1); % w(L/2,t)
96
97   % saving data for drawing
98   if mod(j-1,nt/Ttr)==0;
99   w_3D(1+(j-1)*Ttr/nt,:)=w(:,j+1)';
100  v_3D(1+(j-1)*Ttr/nt,:)=v(:,j+1)';
101  end
102
103  end
104  w_3D(1,:)=w(:,1)';
105  v_3D(1,:)=v(:,1)';
106  for i=1:nx
107  t_3D(:,i)=w_3D(:,i);
108  t_3D(:,nx+i)=v_3D(:,i);
109  end
110  % to reduce the nodes of original nx
111  for i=1:2*nx-1
112  if mod(i,2)==0
113  revise_free(:,i/2)=t_3D(:,i);
114  end
115  end
116
117
118
119
120  %******************************************
121  %         With contorl
122  %******************************************
123  nx=50;
124  nt=8*10^4;
125
126  tmax=50;
127  L=10;
128  Ttr=80;
129
130  dx=L/(nx-1);
131  dt=tmax/(nt-1);
132
133  w=zeros(nx,nt);v=w;
134  u=zeros(nt,1);
135  w_3D=zeros(Ttr,nx);v_3D=w_3D;
136  t_3D=zeros(Ttr,2*nx-1);
137  revise_control=zeros(Ttr,nx-1);
138  u_2D=zeros(Ttr1,1);
139  % initial condition
140  for i=1:nx
141  w(i,1)=( nx-i )*dx*0.3;
142
143  end
144
145  w(:,2)=w(:,1);
146  w(:,3)=w(:,1);
147
148  for i=1:nx
```

```
149    v(i,1)=( i-1)*dx*0.3;
150    end
151    v(:,2)=v(:,1);
152    v(:,3)=v(:,1);
153
154    k=1.25*10^3;
155    alpha=1*10^0;
156    beta=1*10^2;
157    kp=37.5;
158
159
160
161    wl_control=zeros(nt,1);vl_control=wl_control;
162    wl_control(1)=w(1,1);wl_control(2)=wl_control(1);
163    vl_control(1)=v(nx,1);vl_control(2)=vl_control(1);
164    w_L_control=zeros(nt,1);
165    w_L_control(1)=w(nx,1);
166    w_L_control(2)=w(nx,2);
167    %main cycle
168
169    for j=3:nt-1
170
171    for i=3:nx-2
172
173    wxxxx=( w(i+2,j)-4*w(i+1,j)+6*w(i,j)-4*w(i-1,j)+w(i-2,j) )/dx^4;
174    dw=( w(i,j)-w(i,j-1) )/dt;
175    vxxxx=( v(i+2,j)-4*v(i+1,j)+6*v(i,j)-4*v(i-1,j)+v(i-2,j) )/dx^4;
176    dv=( v(i,j)-v(i,j-1) )/dt;
177
178    w(i,j+1)=2*w(i,j)-w(i,j-1)+( -EI*wxxxx-gamma1*dw )*dt^2/rhoA;
179    v(i,j+1)=2*v(i,j)-v(i,j-1)+( -EI*vxxxx-gamma1*dv )*dt^2/rhoA;
180    end
181
182    w(1,j+1)=3*w(3,j+1)-2*w(4,j+1);
183    w(2,j+1)=2*w(3,j+1)-w(4,j+1);
184    v(nx,j+1)=3*v(nx-2,j+1)-2*v(nx-3,j+1);
185    v(nx-1,j+1)=2*v(nx-2,j+1)-v(nx-3,j+1);
186
187
188    wxxxl=( w(nx,j)-3*w(nx-1,j)+3*w(nx-2,j)-w(nx-3,j))/dx^3;
189
190    vxxxl=( v(4,j)-3*v(3,j)+3*v(2,j)-v(1,j) )/dx^3;
191    S2(j)=wxxxl-vxxxl;
192
193    w(nx,j+1)=2*w(nx,j)-w(nx,j-1)+( EI*wxxxl-EI*vxxxl+u(j-1) )*dt^2/m;
194    v(1,j+1)=w(nx,j+1);
195
196    w(nx-1,j+1)=w(nx,j+1);
197    v(2,j+1)=v(1,j+1);
198
199    wl_control(j)=w(1,j+1);
200    vl_control(j)=v(nx,j+1);
201    w_L_control(j)=w(nx,j+1);
202    if mod(j-1,nt/Ttr)==0;
203    w_3D(1+(j-1)*Ttr/nt,:)=w(:,j+1)';
204    v_3D(1+(j-1)*Ttr/nt,:)=v(:,j+1)';
205    end
206    dwl=( w(nx,j+1)-w(nx,j) )/dt;
207    S(j)=alpha/beta*w(nx,j+1)+dwl;
```

```
208   u(j)=-k*S(j)-alpha/beta*m*dwl-kp*w(nx,j+1);
209
210   if mod(j-1,nt/Ttr1)==0
211   u_2D(1+(j-1)*Ttr1/nt)=u(j);
212   end
213   end
214   w_3D(1,:)=w(:,1)';
215   v_3D(1,:)=v(:,1)';
216   wl_2D(1)=w_3D(1,1);
217   vl_2D(1)=v_3D(1,nx);
218   for i=1:nx
219   t_3D(:,i)=w_3D(:,i);
220   t_3D(:,nx+i)=v_3D(:,i);
221   end
222   % to reduce the nodes of orginal nx
223   for i=1:2*nx-1
224   if mod(i,2)==0
225   revise_control(:,i/2)=t_3D(:,i);
226   end
227   end
228
229
230   t_tr=linspace(0,tmax,Ttr);
231
232   % make a draw
233
234   figure(1);
235   surf(linspace(-L,L,nx-1),t_tr,revise_free);view([60 35]);
236   title('Displacement of the flexible satellite without control');
237   ylabel('t [s]');xlabel('x [m]');zlabel('w(x,t) [m]');
238
239
240   figure(2);
241   surf(linspace(-L,L,nx-1),t_tr,revise_control);view([60 35]);
242   title('Displacement of the flexible satellite with control');
243   ylabel('t [s]');xlabel('x [m]');zlabel('w(x,t) [m]');
244
245   figure(3);
246   subplot(211);
247   plot(linspace(0,tmax,nt),wl_free);
248   xlabel('t [s]');ylabel('w(l,t) [m]');
249   subplot(212);
250   plot(linspace(0,tmax,nt),wl_control);
251   xlabel('t [s]');ylabel('w(l,t) [m]');
252
253   figure(4);
254   subplot(211);
255   plot(linspace(0,tmax,nt),w_L_free);
256   xlabel('t [s]');ylabel('w(l/2,t [m]');
257   subplot(212);
258   plot(linspace(0,tmax,nt),w_L_control);
259   xlabel('t [s]');ylabel('w(l/2,t) [m]');
260
261
262   figure(5);
263   subplot(211);
264   plot(linspace(0,tmax,nt),vl_free);
265   xlabel('t [s]');ylabel('w(0,t) [m]');
266   subplot(212);
267   plot(linspace(0,tmax,nt),vl_control);
268   xlabel('t [s]');ylabel('w(0,t) [m]');
```

References

1. Cubillos XCM, de Souza LCG (2009) Using of H-infinity control method in attitude control system of rigid-flexible satellite. Math Probl Eng 2009
2. Do KD, Pan J (2009) Boundary control of three-dimensional inextensible marine risers. J Sound Vib 327(3–5):299–321
3. Guan P, Liu X-J, Liu J-Z (2005) Adaptive fuzzy sliding mode control for flexible satellite. Eng Appl Artif Intell 18(4):451–459
4. He W, Zhang S, Ge SS (2013) Boundary control of a flexible riser with the application to marine installation. IEEE Trans Ind Electron 60(12):5802–5810
5. Hu Q, Ma G (2005) Variable structure control and active vibration suppression of flexible spacecraft during attitude maneuver. Aerosp Sci Technol 9(4):307–317
6. Hu Q, Ma G (2008) Adaptive variable structure controller for spacecraft vibration reduction. IEEE Trans Aerosp Electron Syst 44(3):861–876
7. Liu J, He W (2018) Distributed parameter modeling and boundary control of flexible manipulators. Springer Singapore, Singapore
8. Meng D, Wang X, Xu W, Liang B (2017) Space robots with flexible appendages: dynamic modeling, coupling measurement, and vibration suppression. J Sound Vib 396:30–50
9. Sabatini M, Palmerini GB, Leonangeli N, Gasbarri P (2014) Analysis and experiments for delay compensation in attitude control of flexible spacecraft. Acta Astronaut 104(1):276–292
10. Xu W, Meng D, Chen Y, Qian H, Xu Y (2014) Dynamics modeling and analysis of a flexible-base space robot for capturing large flexible spacecraft. Multibody System Dynamics 32(3):357–401
11. Yang K-J, Hong K-S, Matsuno F (2005) Energy-based control of axially translating beams: varying tension, varying speed, and disturbance adaptation. IEEE Trans Control Syst Technol 13(6):1045–1054

Chapter 4
Boundary Control for Flexible Satellite with Input Constraint

4.1 Introduction

This chapter considers the vibration control problem of a flexible satellite with the restricted input. Previous studies have considered the stability problem under the condition of input saturation [2–4, 9], which are based on nested saturated input functions. Some researchers present anti-windup controllers for linear system using linear matrix inequalities (LMI) [6, 8]. A general framework for the analysis and control of parabolic partial differential equations (PDE) systems with input constraints is developed in [5]. In [7], boundary output feedback control of a flexible string system with input saturation is proposed. In [1], a simple controller with smooth hyperbolic function for achieving trajectory tracking under the condition of restricted input is presented.

In this chapter, we do further research on studying the vibration control for a flexible satellite modelled as PDEs via backstepping method with input constraint and external disturbance. By using backstepping method, an auxiliary system based on a smooth hyperbolic function and a Nussbaum function is designed to deal with the constrained input and external disturbance. Note that the backstepping technique needs all functions differentiable. Therefore, a smooth function is used to approximate the constraint. However, the derivative of the approximate function makes the design and stability analysis a challenge problem. Compared to the existing work, the main contributions of the chapter include:

(i) A hybrid PDE/ODE model of the flexible satellite system under unknown time-varying boundary disturbance for vibration suppression is derived based on Hamilton's principle.

(ii) Back-stepping method with smooth hyperbolic function and an auxiliary system are used to stabilize the flexible satellite under the condition of constrained input and external disturbance;

(iii) With the proposed boundary control, uniform boundedness of the system is proved via Lyapunov synthesis. The closed loop system state will eventually converge

© Tsinghua University Press 2020
Z. Liu and J. Liu, *PDE Modeling and Boundary Control for Flexible
Mechanical System*, Springer Tracts in Mechanical Engineering,
https://doi.org/10.1007/978-981-15-2596-4_4

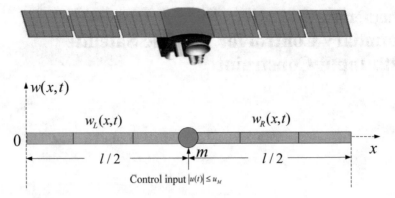

Fig. 4.1 Satellite with two symmetrical flexible solar panels

to a compact set and the control performance of the system is guaranteed by suitably choosing the design parameters.

The rest of this chapter is as follows. In Sect. 4.2, the dynamic model of flexible satellite is given. By using backstepping method, boundary control schemes are proposed to control the deformation of panels in Sect. 4.3. Simulations are carried out to illustrate the performance of the proposed control in Sect. 4.4.

4.2   Problem Formulation

The diagram of a flexible satellite with two solar panels is presented in Fig. 4.1. The satellite dynamic model consists of a centrebody with two symmetrical flexible panels. The centrebody is treated as a rigid body. The external disturbance $d(t)$ is acted on the centrebody of the flexible satellite. The actuator $u(t)$ is imposed at the centrebody to suppress the vibration of the flexible satellite with input constraint and external disturbance. We consider the flexible satellite system (3.10)–(3.14), as shown in Chap. 3, where the governing equations are given as

$$\rho A w_{Ltt}(x, t) + EI w_{Lxxxx}(x, t) + \gamma w_{Lt}(x, t) = 0, \forall x \in [0, l/2] \qquad (4.1)$$

$$\rho A w_{Rtt}(x, t) + EI w_{Rxxxx}(x, t) + \gamma w_{Rt}(x, t) = 0, \forall x \in [l/2, l] \qquad (4.2)$$

The corresponding boundary conditions are given below:

$$w_{Lx}(l/2, t) = w_{Rx}(l/2, t) = 0 \qquad (4.3)$$

$$w_{Lxx}(0, t) = w_{Rxx}(l, t) = 0 \qquad (4.4)$$

$$w_{Lxxx}(0, t) = w_{Rxxx}(l, t) = 0 \tag{4.5}$$

$$mw_{tt}(l/2, t) = EIw_{Lxxx}(l/2, t) - EIw_{Rxxx}(l/2, t) + u(t) + d(t) \tag{4.6}$$

$$w_L(l/2, t) = w_R(l/2, t) = w(l/2, t) \tag{4.7}$$

where ρ represents the density of the panel, $l/2$ represents the length of the panel, A is the cross-sectional area of the panel, m represents the mass of the centrebody. EI and γ are the bending rigidity of the panel and the damping coefficient, respectively. $w_L(x, t)$ and $w_R(x, t)$ are the left displacement and right displacement of the panel, respectively. The displacement of the centrebody is represented by $w(l/2, t)$.

Assumption 4.1 The external disturbancies $d(t)$ is acted on controller placed on the centrebody of the flexible satellite. The unknown boundary disturbance $d(t)$ is bounded. Therefore there are positive constant D that satisfying $|d(t)| \leq D$.

Assumption 4.2 The kinetic energy of the system, is assumed to be bounded $\forall t \in [0, \infty)$, and $\frac{\partial^{q+1}w(x,t)}{\partial t \partial x^q}$ is assumed to be bounded for $t > 0, \forall t \in [0, L), q = 1, 2$.

Assumption 4.3 The potential energy of the system is assumed to be bounded $\forall t \in [0, \infty)$, and $\frac{\partial^p w(x,t)}{\partial x^p}$ is assumed to be bounded for $t > 0, \forall t \in [0, L), p = 2, 3$.

4.3 Boundary Control Design

The control objective of the section is to refrain elastic vibrations of the flexible satellite subject to input constraints, and external disturbances. In this section, a boundary control law $u(t)$ is designed based on the backstepping method and the closed loop stability of the system is analyzed by Lyapunov' direct method.
    To achieve the objective, the input constraint model is described as follows

$$u(t) = u_g(u_0(t)) = u_M \tanh\left(\frac{u_0(t)}{u_M}\right) \tag{4.8}$$

where u_M is a known bound of $u(t)$, and u_0 is the control signal designed in the following process.
    As the usual backstepping method, the following transform of coordinate is made:

$$z_1 = x_1 = w(l/2, t) \tag{4.9}$$

$$z_2 = x_2 - \tau_1 = w_t(l/2, t) - \tau_1 \tag{4.10}$$

$$z_3 = u_g(u_0(t)) - \tau_2 \tag{4.11}$$

where τ_1 and τ_2 are the virtual control laws.

Step 1: We choose the virtual control law τ_1 as

$$\tau_1 = -\frac{\alpha}{\beta}z_1 \tag{4.12}$$

where the parameters $\alpha, \beta > 0$.

A Lyapunov function is designed as

$$V_{b1}(t) = \frac{1}{2}z_1^2 \tag{4.13}$$

The derivative of $V_{b1}(t)$ is

$$\dot{V}_{b1}(t) = z_1\dot{z}_1 = z_1x_2 = z_1(z_2 + \tau_1) \tag{4.14}$$

Substituting Eq. (4.12) into Eq. (4.14), we can obtain

$$\dot{V}_{b1}(t) = -\frac{\alpha}{\beta}z_1^2 + z_1z_2 \tag{4.15}$$

Step 2: The virtual control law τ_2 is designed as

$$\tau_2 = -c_1z_2 - \frac{z_1}{\beta} + m\dot{\tau}_1 - D\tanh\left(\frac{z_2}{\varepsilon}\right) \tag{4.16}$$

where the parameters $c_1 > 0, \varepsilon > 0$.

Then we put forward a Lyapunov function as

$$V_{b2}(t) = V_{b1}(t) + \frac{1}{2}mz_2^2 \tag{4.17}$$

Noting that Eqs. (4.5), (4.8), (4.17), (4.10), (4.11) and (4.15), the derivative of $V_{b2}(t)$ is

$$\begin{aligned}
\dot{V}_{b2}(t) &= \dot{V}_{b1}(t) + \beta mz_2\dot{z}_2 = -\frac{\alpha}{\beta}z_1^2 + z_1z_2 + z_2\beta\,(m\dot{x}_2 - m\dot{\tau}_1)\\
&= -\frac{\alpha}{\beta}z_1^2 + z_1z_2 + z_2\beta\,(EIw_{Lxxx}(l/2, t) - EIw_{Rxxx}(l/2, t) + u(t) + d(t) - m\dot{\tau}_1)\\
&= -\frac{\alpha}{\beta}z_1^2 + z_1z_2 + z_2\beta\,(EIw_{Lxxx}(l/2, t) - EIw_{Rxxx}(l/2, t) + z_3 + \tau_2 + d(t) - m\dot{\tau}_1)
\end{aligned} \tag{4.18}$$

Substituting Eq. (4.16) into Eq. (4.18), we can obtain

$$\dot{V}_{b2}(t) = -\frac{\alpha}{\beta}z_1^2 - c_1\beta z_2^2 + \beta z_2 z_3 + \beta z_2 \left(EIw_{Lxxx}(l/2, t) - EIw_{Rxxx}(l/2, t)\right)$$
$$+ \beta z_2 \left(d(t) - D\tanh\left(\frac{z_2}{\varepsilon}\right)\right) \tag{4.19}$$

According to Lemma 2.7, we can get

$$0 \le D\beta|z_2| - D\beta z_2 \tanh\left(\frac{z_2}{\varepsilon}\right) \le D\beta\mu\varepsilon \tag{4.20}$$

And then

$$-D\beta z_2 \tanh\left(\frac{z_2}{\varepsilon}\right) + \beta z_2 d(t) \le D\beta\mu\varepsilon - D\beta|z_2| + z_2\beta d(t) \le D\beta\mu\varepsilon \tag{4.21}$$

Therefore

$$\dot{V}_{b2}(t) \le -\frac{\alpha}{\beta}z_1^2 - c_1\beta z_2^2 + \beta z_2 z_3 + \beta z_2 \left(EIw_{Lxxx}(l/2, t) - EIw_{Rxxx}(l/2, t)\right) + D\beta\mu\varepsilon \tag{4.22}$$

Substituting Eqs. (4.9), (4.10), (4.12) into Eq. (4.16), we can obtain

$$\tau_2 = -c_1\left(x_2 + \frac{\alpha}{\beta}x_1\right) - \frac{1}{\beta}x_1 - m\frac{\alpha}{\beta}x_2 - D\tanh\left(\frac{x_2 + \frac{\alpha}{\beta}x_1}{\varepsilon}\right) \tag{4.23}$$

Obviously, τ_2 is the function of x_1, x_2. Therefore,

$$m\dot{\tau}_2 = m\frac{\partial\tau_2}{\partial x_1}x_2 + \frac{\partial\tau_2}{\partial x_2}\left(EIw_{Lxxx}(l/2, t) - EIw_{Rxxx}(l/2, t) + u(t) + d(t)\right) \tag{4.24}$$

We project an auxiliary system as

$$\dot{u}_0(t) = -bu_0(t) + v \tag{4.25}$$

where the parameter $b > 0$, v is the auxiliary control law.
Considering Eqs. (4.11) and (4.25), we can get

$$m\dot{z}_3 = m\frac{\partial u_g\left(u_0(t)\right)}{\partial u_0(t)}\left(-bu_0(t) + v\right) - m\dot{\tau}_2$$
$$= m\zeta\left(-bu_0(t) + v\right) - m\dot{\tau}_2 \tag{4.26}$$

where

$$\zeta = \frac{\partial u_g\left(u_0(t)\right)}{\partial u_0(t)} = \frac{4}{\left(e^{u_0(t)/u_M} + e^{-u_0(t)/u_M}\right)^2} > 0 \tag{4.27}$$

Noting that ζ is varying, which makes the design and analysis difficult. There-fore, a Nussbaum function $N(\chi)$ is used to handle this problem, which satisfies the following properties,

$$\lim_{x \to \pm\infty} \sup \frac{1}{k} \int_0^k N(s)ds = \infty \tag{4.28}$$

$$\lim_{x \to \pm\infty} \inf \frac{1}{k} \int_0^k N(s)ds = -\infty \tag{4.29}$$

The auxiliary control law v is designed as

$$v = N(\chi)\bar{v} \tag{4.30}$$

Then we define a Nussbaum function $N(\chi)$ as

$$N(\chi) = \chi^2 \cos(\chi), \quad \dot{\chi} = \gamma_\chi m z_3 \bar{v} \tag{4.31}$$

where γ_χ is a positive real design parameter.

The $\bar{v}$ is designed as

$$\bar{v} = -\frac{c_2}{m}z_3 + \frac{1}{m}\frac{\partial \tau_2}{\partial x_2}u(t) + \frac{\partial \tau_2}{\partial x_1}x_2 + \zeta b u_0(t) - \eta \left(\frac{\partial \tau_2}{\partial x_2}\right)z_3$$
$$-\frac{\beta}{m}z_2 + \frac{1}{m}\frac{\partial \tau_2}{\partial x_2}\left(EIw_{Lxxx}(l/2, t) - EIw_{Rxxx}(l/2, t)\right) \tag{4.32}$$

where the parameter $c_2 > 0$.

Combing Eqs. (4.26) and (4.32), we can obtain

$$m\dot{z}_3 + m\bar{v} = m\zeta v - \frac{\partial \tau_2}{\partial x_2}d(t) - c_2 z_3 - m\eta \left(\frac{\partial \tau_2}{\partial x_2}\right)^2 z_3 - \beta z_2 \tag{4.33}$$

Step 3: A Lyapunov function candidate is designed as

$$V_b(t) = V_{b2}(t) + \frac{1}{2}\beta m z_3^2 \tag{4.34}$$

The derivative of $V_b(t)$ is

$$\dot{V}_b(t) = \dot{V}_{b2}(t) + m z_3 \dot{z}_3$$
$$\leq -\frac{\alpha}{\beta}z_1^2 - c_1\beta z_2^2 + \beta z_2 \left(EIw_{Lxxx}(l/2, t) - EIw_{Rxxx}(l/2, t)\right)$$
$$+ \beta z_2 z_3 + D\beta \mu \varepsilon + m z_3 \dot{z}_3$$

$$\leq -\frac{\alpha}{\beta}z_1^2 - c_1\beta z_2^2 + \beta z_2 \left(EIw_{Lxxx}(l/2, t) - EIw_{Rxxx}(l/2, t)\right)$$

$$+ \beta z_2 z_3 + z_3 \left(m\dot{z}_3 + m\bar{v}\right) - mz_3\bar{v} + D\beta\mu\varepsilon \tag{4.35}$$

Substituting Eqs. (4.30), (4.33) into Eq. (4.35), we have

$$\dot{V}_b(t) \leq -\frac{\alpha}{\beta}z_1^2 - c_1\beta z_2^2 - c_2 z_3^2 + \beta z_2 \left(EIw_{Lxxx}(l/2, t) - EIw_{Rxxx}(l/2, t)\right) + D\beta\mu\varepsilon$$

$$- z_3\frac{\partial\tau_2}{\partial x_2}d(t) - m\eta\left(\frac{\partial\tau_2}{\partial x_2}\right)^2 z_3^2 + (\zeta N(\chi) - 1)mz_3\bar{v} \tag{4.36}$$

Noting that

$$\frac{1}{4\eta m}d^2(t) + m\eta\left(\frac{\partial\tau_2}{\partial x_2}\right)^2 z_3^2 \geq -z_3\frac{\partial\tau_2}{\partial x_2}d(t) \tag{4.37}$$

$$- z_3\frac{\partial\tau_2}{\partial x_2}d(t) - m\eta\left(\frac{\partial\tau_2}{\partial x_2}\right)^2 z_3^2 \leq \frac{1}{4\eta m}d^2(t) \tag{4.38}$$

Therefore,

$$\dot{V}_b(t) \leq -\frac{\alpha}{\beta}z_1^2 - c_1\beta z_2^2 - c_2 z_3^2 + \beta z_2 \left(EIw_{Lxxx}(l/2, t) - EIw_{Rxxx}(l/2, t)\right) + D\beta\mu\varepsilon$$

$$+ \frac{1}{4\eta m}D^2 + (\zeta N(\chi) - 1)mz_3\bar{v}$$

$$\leq -\frac{\alpha}{\beta}z_1^2 - c_1\beta z_2^2 - c_2 z_3^2 + \beta z_2 \left(EIw_{Lxxx}(l/2, t) - EIw_{Rxxx}(l/2, t)\right) + D\beta\mu\varepsilon$$

$$+ \frac{1}{4\eta m}D^2 + \frac{1}{\gamma_\chi}(\zeta N(\chi) - 1)\dot{\chi} \tag{4.39}$$

With the above steps, we are ready to present the following theorem of the closed-loop system.

Theorem 4.1 *With the proposed control law (4.12), (4.16), (4.30) and (4.31), the following properties hold:*

(1) The closed-loop system is uniformly bounded, that is $|w_L(x, t)| \leq C_L$ and $|w_R(x, t)| \leq C_R$, where $C_L = C_R = \sqrt{L^3 V(t)/4\beta EI\mu_2}$.

(2) The control input is bounded, and its bound is $u(t) = u_M \tanh(u_0(t)/u_M) \leq u_M$.

Proof Considering the Lyapunov function

$$V(t) = V_1(t) + V_2(t) + V_b(t) \tag{4.40}$$

where

$$V_1(t) = \frac{\beta}{2}\rho A \int_0^{l/2} [w_{Lt}(x,t)]^2 dx + \frac{\beta}{2}EI \int_0^{l/2} [w_{Lxx}(x,t)]^2 dx$$
$$+ \frac{\alpha}{2}\gamma \int_0^{l/2} [w_L(x,t)]^2 dx + \frac{\beta}{2}\rho A \int_{l/2}^l [w_{Rt}(x,t)]^2 dx$$
$$+ \frac{\beta}{2}EI \int_{l/2}^l [w_{Rxx}(x,t)]^2 dx + \frac{\alpha}{2}\gamma \int_{l/2}^l [w_R(x,t)]^2 dx \qquad (4.41)$$

$$V_2(t) = \alpha\rho A \int_0^{l/2} w_{Lt}(x,t)w_L(x,t)dx + \alpha\rho A \int_{l/2}^l w_{Rt}(x,t)w_R(x,t)dx \quad (4.42)$$

The Lyapunov function $V_1(t)$ is bounded as

$$V_1(t) \geq \phi_1 \left[\int_0^{l/2} \left([w_{Lt}(x,t)]^2 + [w_L(x,t)]^2 \right)dx + \int_{l/2}^l \left([w_{Rt}(x,t)]^2 + [w_R(x,t)]^2 \right)dx \right]$$
$$(4.43)$$

where $\phi_1 = \min\left(\frac{\beta\rho A}{2}, \frac{\alpha\gamma}{2}\right) > 0$.

The Lyapunov function $V_2(t)$ is bounded as

$$|V_2(t)| \leq \alpha\rho A \left[\int_0^{l/2} \left([w_{Lt}(x,t)]^2 + [w_L(x,t)]^2 \right)dx + \int_{l/2}^l \left([w_{Rt}(x,t)]^2 + [w_R(x,t)]^2 \right)dx \right]$$
$$\leq \phi_2 V_1(t) \qquad (4.44)$$

where $\phi_2 = \frac{\alpha\rho A}{2\phi_1}$.

Considering $\phi_1 > \frac{1}{2}\alpha\rho A$, we can have

$$0 \leq \phi_4 V_1(t) \leq V_1(t) + V_2(t) \leq \phi_3 V_1(t) \qquad (4.45)$$

where $\phi_3 = 1 + \phi_2 > 1$, $\phi_4 = 1 - \phi_2 < 1$.

Considering (4.40), we can obtain

$$\mu_2 (V_1(t) + V_b(t)) \leq V(t) \leq \mu_1 (V_1(t) + V_b(t)) \qquad (4.46)$$

where $\mu_1 = \max(\phi_3, 1) = \phi_3$, $\mu_2 = \min(\phi_4, 1) = \phi_4$.

The derivative of (4.40), we have

$$\dot{V}(t) = \dot{V}_1(t) + \dot{V}_2(t) + \dot{V}_b(t) \qquad (4.47)$$

where $\dot{V}_1(t)$ is given as

$$\dot{V}_1(t) = \beta \rho A \int_0^{l/2} w_{Lt}(x,t)w_{Ltt}(x,t)dx + \beta EI \int_0^{l/2} w_{Lxx}(x,t)w_{Lxxt}(x,t)dx$$

$$+ \alpha \gamma \int_0^{l/2} w_L(x,t)w_{Lt}(x,t)dx + \beta \rho A \int_{l/2}^l w_{Rt}(x,t)w_{Rtt}(x,t)dx$$

$$+ \beta EI \int_{l/2}^l w_{Rxx}(x,t)w_{Rxxt}(x,t)dx + \alpha \gamma \int_{l/2}^l w_R(x,t)w_{Rt}(x,t)dx$$

$$(4.48)$$

Substituting Eq. (4.1) into Eq. (4.48), and using the boundary conditions, we can obtain

$$\dot{V}_1(t) = -\beta \gamma \int_0^{l/2} [w_{Lt}(x,t)]^2 dx - \beta \gamma \int_{l/2}^l [w_{Rt}(x,t)]^2 dx$$

$$+ \alpha \gamma \int_0^{l/2} w_L(x,t)w_{Lt}(x,t)dx + \alpha \gamma \int_{l/2}^l w_R(x,t)w_{Rt}(x,t)dx$$

$$- \beta w_t(l/2,t)\left(EIw_{Lxxx}(l/2,t) - EIw_{Rxxx}(l/2,t)\right) \qquad (4.49)$$

where $\dot{V}_2(t)$ is

$$\dot{V}_2(t) = \alpha \rho A \int_0^{l/2} w_{Ltt}(x,t)w_L(x,t)dx + \alpha \rho A \int_0^{l/2} [w_{Lt}(x,t)]^2 dx$$

$$+ \alpha \rho A \int_{l/2}^l w_{Rtt}(x,t)w_R(x,t)dx + \alpha \rho A \int_{l/2}^l [w_{Rt}(x,t)]^2 dx \qquad (4.50)$$

Substituting Eqs. (4.1) and (4.2) into Eq. (4.50), and using the boundary conditions, we can obtain

$$\dot{V}_2(t) = -\alpha w(l/2,t)\left(EIw_{Lxxx}(l/2,t) - EIw_{Rxxx}(l/2,t)\right) - \alpha EI \int_0^{l/2} [w_{Lxx}(x,t)]^2 dx$$

$$- \alpha EI \int_{l/2}^l [w_{Rxx}(x,t)]^2 dx + \alpha \rho A \int_0^{l/2} [w_{Lt}(x,t)]^2 dx + \alpha \rho A \int_{l/2}^l [w_{Rt}(x,t)]^2 dx$$

$$- \alpha \gamma \int_0^{l/2} w_L(x,t)w_{Lt}(x,t)dx - \alpha \gamma \int_{l/2}^l w_R(x,t)w_{Rt}(x,t)dx \qquad (4.51)$$

Substituting Eqs. (4.39), (4.49) and (4.51) into Eq. (4.47), we can obtain

$$\dot{V}(t) = -\frac{\alpha}{\beta}z_1^2 - c_1\beta z_2^2 - c_2 z_3^2 - (\gamma \beta - \alpha \rho A) \int_0^{l/2} [w_{Lt}(x,t)]^2 dx$$

$$- (\gamma \beta - \alpha \rho A) \int_{l/2}^l [w_{Rt}(x,t)]^2 dx - \alpha EI \int_0^{l/2} [w_{Lxx}(x,t)]^2 dx$$

$$-\alpha EI \int_{l/2}^{l} [w_{Rxx}(x,t)]^2 dx + \frac{1}{\gamma_\chi} (\zeta N(\chi) - 1) \dot{\chi} + D\beta\mu\varepsilon + \frac{1}{4\eta m} D^2$$

$$(4.52)$$

According to Lemma 2.6, the following inequalities can be obtained

$$-\varphi_1 l [w(l/2, t)]^2 \leq -\varphi_1 \int_0^{l/2} [w_L(x,t)]^2 dx + \varphi_1 l^4 \int_0^{l/2} [w_{Lxx}(x,t)]^2 dx \quad (4.53)$$

$$-\varphi_2 l [w(l/2, t)]^2 \leq -\varphi_2 \int_{l/2}^{l} [w_R(x,t)]^2 dx + \varphi_2 l^4 \int_{l/2}^{l} [w_{Rxx}(x,t)]^2 dx \quad (4.54)$$

where φ_1 and φ_2 are positive constants. Then we can obtain

$$\dot{V}(t) \leq -\left(\frac{\alpha}{\beta} - \varphi_1 l - \varphi_2 l\right) z_1^2 - c_1\beta z_2^2 - c_2 z_3^2 - (\gamma\beta - \alpha\rho A) \int_0^{l/2} [w_{Lt}(x,t)]^2 dx$$

$$- (\gamma\beta - \alpha\rho A) \int_{l/2}^{l} [w_{Rt}(x,t)]^2 dx - \left(\alpha EI - \varphi_1 l^4\right) \int_0^{l/2} [w_{Lxx}(x,t)]^2 dx$$

$$- \left(\alpha EI - \varphi_2 l^4\right) \int_{l/2}^{l} [w_{Rxx}(x,t)]^2 dx - \varphi_1 \int_0^{l/2} [w_L(x,t)]^2 dx$$

$$- \varphi_2 \int_{l/2}^{l} [w_R(x,t)]^2 dx + \frac{1}{\gamma_\chi} (\zeta N(\chi) - 1) \dot{\chi} + D\beta\mu\varepsilon + \frac{1}{4\eta m} D^2$$

$$(4.55)$$

Choosing the parameters c_1, c_2, α, β, φ_1, φ_2 and l to satisfy the following conditions:

$$A_1 = c_1\beta > 0$$
$$A_2 = c_2 > 0$$
$$A_3 = \gamma\beta - \alpha\rho A > 0$$
$$A_4 = \alpha EI - \varphi_1 l^4 > 0$$
$$A_5 = \alpha EI - \varphi_2 l^4 > 0$$
$$A_6 = \varphi_1 > 0$$
$$A_7 = \varphi_2 > 0$$
$$A_8 = \frac{\alpha}{\beta} - \varphi_1 l - \varphi_2 l > 0$$

Therefore

$$\dot{V}(t) \leq -\mu_3 (V_1(t) + V_b(t)) + \frac{1}{\gamma_\chi} (\zeta N(\chi) - 1) \dot{\chi} + D\beta\mu\varepsilon + \frac{1}{4\eta m} D^2 \quad (4.56)$$

where $\mu_3 = 2 \min \left(\frac{A_1}{m\beta}, \frac{A_2}{m}, \frac{A_3}{\beta\rho A}, \frac{A_4}{\beta EI}, \frac{A_5}{\beta EI}, \frac{A_6}{\alpha\gamma}, \frac{A_7}{\alpha\gamma}, A_8 \right) > 0.$

Combining Eqs. (4.46) and (4.56), we can obtain

$$\dot{V}(t) \leq -\mu V(t) + \frac{1}{\gamma_\chi} (\zeta N(\chi) - 1) \dot{\chi} + D\beta\mu\varepsilon + \frac{1}{4\eta m} D^2 \qquad (4.57)$$

where $\mu = \mu_3/\mu_1 > 0$, $\mu \geq \max \left\{ \frac{1}{V(t)} \left(\frac{1}{\gamma_\chi} (\zeta N(\chi) - 1) \dot{\chi} + D\beta\mu\varepsilon + \frac{1}{4\eta m} D^2 \right) \right\}$.

Then integrating Eq. (4.57), we can obtain

$$V(t) \leq V(0)e^{-\mu t} + \frac{1}{\mu} \left(D\beta\mu\varepsilon + \frac{1}{4\eta m} D^2 \right) \left(1 - e^{-\mu t} \right)$$

$$+ \frac{e^{-\mu t}}{\gamma_\chi} \int_0^t (\zeta N(\chi) - 1) \dot{\chi} e^{-\mu t} d\tau \qquad (4.58)$$

According to Lemma 2.8, we can draw a conclusion that $V(t)$ and χ are bounded on $[0, t)$. We can further obtain that z_1, z_2, z_3, $w(x, t)$, and $w_t(x, t)$ are all bounded. Then based on Assumptions 2 and 3, and the following properties can be hold:

(i) $\left| u_g(u_0) \right| = u_M \left| \tanh \left(\frac{u_0(t)}{u_M} \right) \right| \leq u_M$

(ii) $\left| \frac{\partial u_g(u_0)}{\partial u_0} \right| = \left| \frac{4}{(e^{u_0/u_M} + e^{-u_0/u_M})^2} \right| \leq 1$

(iii) $\left| \frac{\partial u_g(u_0)}{\partial u_0} u_0 \right| = \left| \frac{4u_0}{(e^{u_0/u_M} + e^{-u_0/u_M})^2} \right| \leq \frac{u_M}{2}$

We further obtain that $\bar{v}$, v, and $u_0(t)$ are all bounded. Moreover, combining the following inequality $\int_0^L [w_{xx}(x, t)]^2 dx \geq \frac{1}{L^2} \int_0^L [w_x(x, t)]^2 dx \geq \frac{1}{L^3} [w(x, t)]^2$ we have

$$\frac{4\beta EI}{l^3} [w_L(x, t)]^2 \leq \frac{2\beta EI}{l^2} \int_0^{l/2} [w_{Lx}(x, t)]^2 dx \leq \frac{\beta EI}{2} \int_0^{l/2} [w_{Lxx}(x, t)]^2 dx$$

$$\leq V_1(t) \leq V_1(t) + V_b(t) \leq \frac{V(t)}{\mu_2} \qquad (4.59)$$

We can further obtain

$$|w_L(x, t)| \leq \sqrt{\frac{l^3 V(t)}{4\beta EI \mu_2}} \qquad (4.60)$$

Similarly, we have

$$|w_R(x, t)| \leq \sqrt{\frac{L^3 V(t)}{4\beta EI \mu_2}} \qquad (4.61)$$

Table 4.1 Parameters of the flexible satellite

Parameter	Description	Value
$l/2$	The length of the panel	$10\,\mathrm{m}$
ρ	Uniform mass per unit length of the panel	$2.7 \times 10^2\,\mathrm{kg/m^3}$
m	The mass of the centrebody	$100\,\mathrm{kg}$
EI	The bending rigidity of the panel	$11.95 \times 10^3\,\mathrm{Nm^3}$
γ	The coefficient of viscous damping	$10\,\mathrm{kg/(ms)}$
A	The cross-sectional area of the panel	$0.12\,\mathrm{m^2}$

4.4   Simulation

For the purpose of illustrating the system performance, we use the finite difference method to show the simulation results. Parameters of the flexible satellite are given as in the Table 4.1.

The initial conditions of the flexible panel are given as:

$$w_L(x, 0) = -0.3x,$$
$$w_R(x, 0) = 0.3x.$$

The external disturbance is given as

$$d(t) = 0.1 \sin (2\pi t) \, (N)$$

The control input is constrained as

$$|u(t)| \leq u_M = 1200N$$

The control parameters are chosen as $c_1 = 750, c_2 = 40, \alpha = 10, \beta = 50, b = 5,$ $\varepsilon = 0.1, \eta = 0.01, \varphi_1 = 0.009, \varphi_2 = 0.01.$

Simulation results without control (i.e., $u(t) = 0$) are given in Fig. 4.2. For the results presented here, we can observe that there are large vibrations along the two panels.

The dynamic responses with control are displayed in Figs. 4.3, 4.4, 4.5, 4.6 and 4.7. Figure 4.3 demonstrates the displacement of the flexible satellite with the proposed control. Figure 4.4 shows the left boundary displacement $w(0, t)$ of the flexible satellite. The displacement $w(l/2, t)$ of the centrebody is demonstrated in Fig. 4.5. In Fig. 4.6, the right boundary displacement $w(l, t)$ of the flexible satellite is demonstrated. Control input is shown in Fig. 4.7.

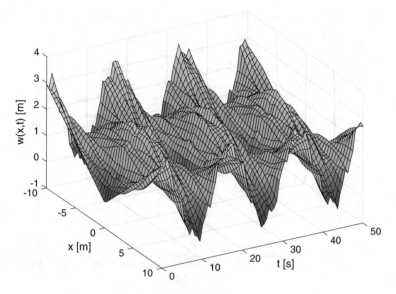

Fig. 4.2 Deformation of flexible satellite without control

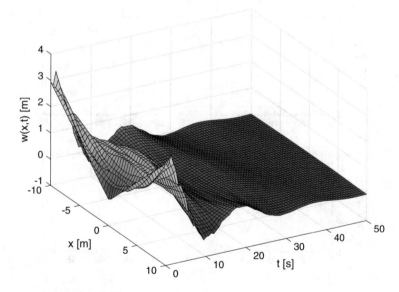

Fig. 4.3 Deformation of flexible satellite with proposed control

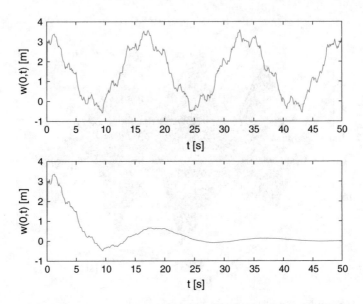

Fig. 4.4 Boundary displacement $w(0, t)$ of flexible satellite: without control and with control

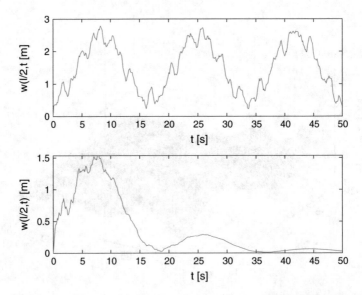

Fig. 4.5 Displacement of center body $w(l/2, t)$ of flexible satellite: without control and with control

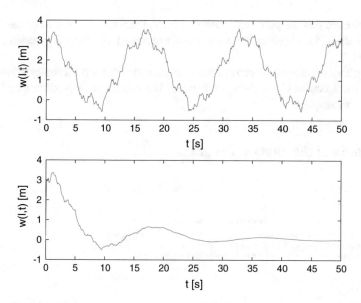

Fig. 4.6 Boundary displacement $w(l, t)$ of flexible satellite: without control and with control

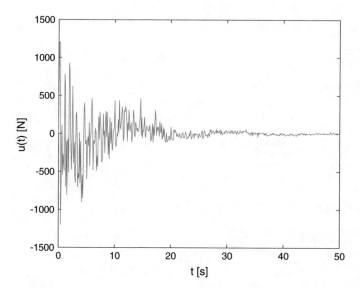

Fig. 4.7 Control input

We can see that the proposed controller can suppress the vibrations of the flexible satellite after 25 s, which means a good performance of vibration suppressing can be acquired.

From above analysis, we can draw a conclusion that the control strategy developed in this work can stabilize the system under the condition of input constraint and external disturbance.

Appendix 1: Simulation Program

```
1   clc;
2   close all;
3   clear all;
4   %%
5   %***********************************
6   %        Expoetial stablization of a
7   %     flexible satellite withou control
8   %***********************************
9
10  nx=50;
11  nt=8*10^4;
12
13  tmax=50;
14  L=10;
15  Ttr=80;
16  tmax=50;
17
18  Ttr1=5*10^2;
19  Ttr2=5*10^2;
20
21  dx=L/(nx-1);
22  dt=tmax/(nt-1);
23
24  w=zeros(nx,nt);v=w;
25  u=zeros(nt,1);
26  w_3D=zeros(Ttr,nx);v_3D=w_3D;
27  t_3D=zeros(Ttr,2*nx-1);
28  revise_free=zeros(Ttr,nx-1);
29
30  m=1;
31  A=0.12;
32  E=6.894*10^10;
33
34  rho=2.7*10^3;
35  I=1.734*10^-7;
36  gamma1=0.005;
37  rhoA=rho*A;
38  EI=E*I;
39
40  % initial condition
41  for i=1:nx
42  w(i,1)=( nx-i  )*dx*0.3;
43  end
44
45  w(:,2)=w(:,1);
46  w(:,3)=w(:,1);
47
48  for i=1:nx
49  v(i,1)=( i-1)*dx*0.3;
```

```
50  end
51  v(:,2)=v(:,1);
52  v(:,3)=v(:,1);
53
54  wl_free=zeros(nt,1);vl_free=wl_free;
55  wl_free(1)=w(1,1);wl_free(2)=wl_free(1);
56  vl_free(1)=v(nx,1);vl_free(2)=v(1);
57  w_L_free=zeros(nt,1);
58  w_L_free(1)=w(nx,1);
59  w_L_free(2)=w(nx,1);
60  %main cycle
61
62  for j=3:nt-1
63
64  for i=3:nx-2
65
66  wxxxx=( w(i+2,j)-4*w(i+1,j)+6*w(i,j)-4*w(i-1,j)+w(i-2,j) )/dx^4;
67  dw=( w(i,j)-w(i,j-1) )/dt;
68  vxxxx=( v(i+2,j)-4*v(i+1,j)+6*v(i,j)-4*v(i-1,j)+v(i-2,j) )/dx^4;
69  dv=( v(i,j)-v(i,j-1) )/dt;
70
71  w(i,j+1)=2*w(i,j)-w(i,j-1)+( -EI*wxxxx-gamma1*dw )*dt^2/rhoA;
72  v(i,j+1)=2*v(i,j)-v(i,j-1)+( -EI*vxxxx-gamma1*dv )*dt^2/rhoA;
73  end
74
75  w(1,j+1)=3*w(3,j+1)-2*w(4,j+1);
76  w(2,j+1)=2*w(3,j+1)-w(4,j+1);
77  v(nx,j+1)=3*v(nx-2,j+1)-2*v(nx-3,j+1);
78  v(nx-1,j+1)=2*v(nx-2,j+1)-v(nx-3,j+1);
79
80
81  wxxxl=( w(nx,j)-3*w(nx-1,j)+3*w(nx-2,j)-w(nx-3,j))/dx^3;
82
83  vxxxl=( v(4,j)-3*v(3,j)+3*v(2,j)-v(1,j) )/dx^3;
84  S(j)=wxxxl-vxxxl;
85  Q(j)=wxxxl;
86  R(j)=vxxxl;
87  w(nx,j+1)=2*w(nx,j)-w(nx,j-1)+( EI*wxxxl-EI*vxxxl+u(j-1) )*dt^2/m;
88  v(1,j+1)=w(nx,j+1);
89
90  w(nx-1,j+1)=w(nx,j+1);
91  v(2,j+1)=v(1,j+1);
92
93  wl_free(j)=w(1,j+1);
94  vl_free(j)=v(nx,j+1);
95  w_L_free(j)=w(nx,j+1); % w(L/2,t)
96  if mod(j-1,nt/Ttr)==0;
97  w_3D(1+(j-1)*Ttr/nt,:)=w(:,j+1)';
98  v_3D(1+(j-1)*Ttr/nt,:)=v(:,j+1)';
99  end
100
101  end
102  w_3D(1,:)=w(:,1)';
103  v_3D(1,:)=v(:,1)';
104  for i=1:nx
105  t_3D(:,i)=w_3D(:,i);
106  t_3D(:,nx+i)=v_3D(:,i);
107  end
108  % to reduce the nodes of original nx
109  for i=1:2*nx-1
110  if mod(i,2)==0
111  revise_free(:,i/2)=t_3D(:,i);
112  end
```

```
113  end
114
115
116  t_tr=linspace(0,tmax,Ttr);
117
118  %*****************************************
119  %           with contorl
120  %*****************************************
121
122  nx=50;
123  nt=8*10^4;
124
125  tmax=50;
126  L=10;
127  Ttr=80;
128
129  dx=L/(nx-1);
130  dt=tmax/(nt-1);
131
132  w=zeros(nx,nt);v=w;
133  u=zeros(nt,1);
134  w_3D=zeros(Ttr,nx);v_3D=w_3D;
135  t_3D=zeros(Ttr,2*nx-1);
136  revise_control=zeros(Ttr,nx-1);
137  u_2D=zeros(Ttr1,1);
138  % initial condition
139  for i=1:nx
140  w(i,1)=( nx-i )*dx*0.3;
141
142  end
143
144  w(:,2)=w(:,1);
145  w(:,3)=w(:,1);
146
147  for i=1:nx
148  v(i,1)=( i-1)*dx*0.3;
149  end
150  v(:,2)=v(:,1);
151  v(:,3)=v(:,1);
152  wxxx1=0;
153
154  vxxx1=0;
155  w1=0;
156  dw1=0;
157  for j=1:nt
158  ut(j)=0;
159
160  X(j)=0;
161  u0(j)=0;
162  gv(j)=0;
163
164  end
165
166  wl_control=zeros(nt,1);vl_control=wl_control;
167  wl_control(1)=w(1,1);wl_control(2)=wl_control(1);
168  vl_control(1)=v(nx,1);vl_control(2)=vl_control(1);
169  w_L_control=zeros(nt,1);
170  w_L_control(1)=w(nx,1);
171  w_L_control(2)=w(nx,2);
172  %main cycle
176  for j=2:nt-1
177
178  x1=w1;
```

```
179    x2=dw1;
180    beta=1100;
181    alpha=25;
182    c3=40;
183    c1=750;
184    c4=1/beta;
185    uM=1200;
186    c=5;
187
188    z1=x1;
189    dz1=x2;
190    tau1=-alpha*z1*c4;
191    dtau1=-alpha*c4*dz1;
192    z2=x2-tau1;
193
194
195    tau2=-z1*c4+m*dtau1-c1*z2;
196    dtau2_x1=-c4-c1*c4*alpha;
197    dtau2_x2=-c1-alpha*c4*m;
198
199    dg_v=4/(exp(u0(j)/uM)+exp(-u0(j)/uM))^2;
200
201    z3=gv(j)-tau2;
202
203    zeta=m*dtau2_x1*x2+dtau2_x2*(EI*wxxxl-EI*vxxxl+gv(j));
204
205
206    wb=(-c3*z3+zeta+m*c*u0(j)*dg_v-z2*beta)/m;
207    gamax= 1.5000e-13;
208    X(j)=X(j-1)+dt*gamax*m*z3*wb;
209
210    N=X(j)^2*cos(X(j));
211    ow=N*wb;
212    u0(j+1)=u0(j)+dt*(-c*u0(j)+ow);
213
214    gv(j+1)=uM*tanh(u0(j+1)/uM);
215
216
217
218    for i=3:nx-2
219    wxxxx=( w(i+2,j)-4*w(i+1,j)+6*w(i,j)-4*w(i-1,j)+w(i-2,j) )/dx^4;
220    dw=( w(i,j)-w(i,j-1) )/dt;
221    vxxxx=( v(i+2,j)-4*v(i+1,j)+6*v(i,j)-4*v(i-1,j)+v(i-2,j) )/dx^4;
222    dv=( v(i,j)-v(i,j-1) )/dt;
223
224    w(i,j+1)=2*w(i,j)-w(i,j-1)+( -EI*wxxxx-gamma1*dw )*dt^2/rhoA;
225    v(i,j+1)=2*v(i,j)-v(i,j-1)+( -EI*vxxxx-gamma1*dv )*dt^2/rhoA;
226    end
227
228    w(1,j+1)=3*w(3,j+1)-2*w(4,j+1);
229    w(2,j+1)=2*w(3,j+1)-w(4,j+1);
230    v(nx,j+1)=3*v(nx-2,j+1)-2*v(nx-3,j+1);
231    v(nx-1,j+1)=2*v(nx-2,j+1)-v(nx-3,j+1);
232
233    w(nx,j+1)=2*w(nx,j)-w(nx,j-1)+( EI*wxxxl-EI*vxxxl+u(j) )*dt^2/m;
234    v(1,j+1)=w(nx,j+1);
235
236    w(nx-1,j+1)=w(nx,j+1);
237    v(2,j+1)=v(1,j+1);
238
239    wl=w(nx,j+1);
240
241    wxxxl=( w(nx,j+1)-3*w(nx-1,j+1)+3*w(nx-2,j+1)-w(nx-3,j+1))/dx^3;
```

```
242   vxxx1=(  v(4,j+1)-3*v(3,j+1)+3*v(2,j+1)-v(1,j+1)  )/dx^3;
243
244
245   w1_control(j)=w(1,j+1);
246   v1_control(j)=v(nx,j+1);
247   w_L_control(j)=w(nx,j+1);
248   if mod(j-1,nt/Ttr)==0;
249   w_3D(1+(j-1)*Ttr/nt,:)=w(:,j+1)';
250   v_3D(1+(j-1)*Ttr/nt,:)=v(:,j+1)';
251   end
252   dw1=(  w(nx,j+1)-w(nx,j)  )/dt;
253
254
255
256   u(j+1)=gv(j+1);
257
258
259   if mod(j-1,nt/Ttr1)==0
260   u_2D(1+(j-1)*Ttr1/nt)=u(j);
261
262   end
263
264   if mod(j-1,nt/Ttr2)==0
265
266   w1_free_2D(1+(j-1)*Ttr2/nt)=w1_free(j);
267   v1_free_2D(1+(j-1)*Ttr2/nt)=v1_free(j);
268   w_L_free_2D(1+(j-1)*Ttr2/nt)=w_L_free(j);
269
270   w1_control_2D(1+(j-1)*Ttr2/nt)=w1_control(j);
271   v1_control_2D(1+(j-1)*Ttr2/nt)=v1_control(j);
272   w_L_control_2D(1+(j-1)*Ttr2/nt)=w_L_control(j);
273
274   end
275   end
276   w_3D(1,:)=w(:,1)';
277   v_3D(1,:)=v(:,1)';
278   w1_2D(1)=w_3D(1,1);
279   v1_2D(1)=v_3D(1,nx);
280   for i=1:nx
281   t_3D(:,i)=w_3D(:,i);
282   t_3D(:,nx+i)=v_3D(:,i);
283   end
284   % to reduce the nodes of orginal nx
285   for i=1:2*nx-1
286   if mod(i,2)==0
287   revise_control(:,i/2)=t_3D(:,i);
288   end
289   end
290
291
292   t_tr=linspace(0,tmax,Ttr);
293   t_tr1=linspace(0,tmax,Ttr1);
294   t_tr2=linspace(0,tmax,Ttr2);
295
296   figure(1);
297   surf(linspace(-L,L,nx-1),t_tr,revise_free);view([60 35]);
298   % title('Displacement of the panel without control');
299   ylabel('t [s]');xlabel('x [m]');zlabel('w(x,t) [m]');
300
301   figure(2);
302   surf(linspace(-L,L,nx-1),t_tr,revise_control);view([60 35]);
303   % title('Displacement of the  panel with control');
304   ylabel('t [s]');xlabel('x [m]');zlabel('w(x,t) [m]');
```

```
305
306  figure(3);
307  subplot(211);
308  plot(t_tr2,wl_free_2D);
309  % title('Deflection w(0,t) of the panel without control');
310  xlabel('t [s]');ylabel('w(l,t) [m]');
311  subplot(212);
312  plot(t_tr2,wl_control_2D);
313  % title('Deflection w(L,t) of the panel with control');
314  xlabel('t [s]');ylabel('w(l,t) [m]');
315  figure(4);
316  subplot(211);
317  plot(t_tr2,w_L_free_2D);
318  xlabel('t [s]');ylabel('w(l/2,t [m]');
319  % title('Deflection w(L/2.t) of the panel without control');
320  subplot(212);
321  plot(t_tr2,w_L_control_2D);
322  xlabel('t [s]');ylabel('w(l/2,t) [m]');
323  % title('Deflection w(L/2,t) of the panel with control');
324
325  figure(5);
326  subplot(211);
327  plot(t_tr2,vl_free_2D);
328  % title('Deflection w(0,t) of the panel without control');
329  xlabel('t [s]');ylabel('w(0,t) [m]');
330  subplot(212);
331  plot(t_tr2,vl_control_2D);
332  % title('Deflection w(L,t) of the panel with control');
333  xlabel('t [s]');ylabel('w(0,t) [m]');
334
335  figure(6);
336  plot(t_tr1,u_2D);
337  xlabel('t [s]');ylabel('u(t) [N]');
338  % title('Control input');
```

References

1. Ailon A (2010) Simple tracking controllers for autonomous VTOL aircraft with bounded inputs. IEEE Trans Autom Control 55(3):737–743
2. Benzaouia A, Akhrif O, Saydy L (2010) Stabilisation and control synthesis of switching systems subject to actuator saturation. Int J Syst Sci 41(4):397–409
3. Boškovic JD, Li S-M, Mehra RK (2001) Robust adaptive variable structure control of spacecraft under control input saturation. J Guid, Control, Dyn 24(1):14–22
4. Chen Y, Fei S, Zhang K (2014) Stabilisation for switched linear systems with time-varying delay and input saturation. Int J Syst Sci 45(3):532–546
5. El-Farra NH, Armaou A, Christofides PD (2003) Analysis and control of parabolic pde systems with input constraints. Automatica 39(4):715–725
6. Grimm G, Hatfield J, Postlethwaite I, Teel AR, Turner MC, Zaccarian L (2003) Antiwindup for stable linear systems with input saturation: an LMI-based synthesis. IEEE Trans Autom Control 48(9):1509–1525
7. He W, He X, Ge SS (2015) Boundary output feedback control of a flexible string system with input saturation. Nonlinear Dyn, pp 1–18
8. Mulder EF, Kothare MV, Morari M (2001) Multivariable anti-windup controller synthesis using linear matrix inequalities. Automatica 37(9):1407–1416
9. Wen C, Zhou J, Liu Z, Hongye S (2011) Robust adaptive control of uncertain nonlinear systems in the presence of input saturation and external disturbance. EEE Trans Autom Control 56(7):1672–1678

Chapter 5
PDE Modeling and Basic Vibration Control for Flexible Aerial Refueling Hose

5.1 Introduction

As the number of unmanned aerial vehicles (UAVs) increases in modern military missions, autonomous aerial refueling (AAR) has gained substantial attention, and significant research is carried out for the detection, control and guidance of the tanker and the receiver [1, 3, 6, 10, 12]. A hose-drogue aerial refueling system consists of a flexible hose-drogue on a tanker and a probe on a receiver, which is the most universal refueling equipment because of its various advantages such as simple and cheap tanker, multipoint hose-drogue, and no boom operator. However, due to the flexible property of the hose, the deflection of the flexible hose has a significant influence on the dynamics and control performance of the AAR. Therefore, the vibration suppression is a vital research relevant to a flexible aerial refueling hose. In [8], a dynamic model of the variable-length hose-drogue aerial refueling system (HDRS) is built and an integral sliding mode back-stepping controller is proposed for the whipping phenomenon. In [9], the optimal control of an aerial-towed flexible cable system is proposed to account for the bowing of the cable. The control design for a previously developed aerial refueling hose-drogue system during receiver-probe coupling is studied in [7]. These works relate to two types of modeling approaches, which are the elasto-dynamic hose model based on the finite element method (FEM), and the lumped mass hose model with rigid link kinematics. However, the two approaches based on truncated models can cause spillover effects, which result in instability when the control of the system is restricted to a few critical modes [4]. The control order needs to be increased with the number of flexible modes considered to obtain high accuracy of performance. To avoid these problems, the flexible hose is regarded as a distributed parameter system which is infinite dimensional and described by partial differential equations (PDEs). However, it adds unique challenges for the control design.

Moreover, the model of a flexible hose is different from PDE models in the existing studies. A flexible aerial refueling hose system needs consider the horizontal

© Tsinghua University Press 2020
Z. Liu and J. Liu, *PDE Modeling and Boundary Control for Flexible Mechanical System*, Springer Tracts in Mechanical Engineering,
https://doi.org/10.1007/978-981-15-2596-4_5

velocity and gravity. Different from the axially moving string in [2], the hose moves horizontally at an angle, which makes the control problem in this chapter more difficult to handle compared to the previous works.

In this chapter, we investigate the boundary control problem for a flexible aerial refueling hose. The refueling process begins with the tanker deploying the hose. Assuming that the tanker flies horizontally and considering the aerodynamic forces and the gravity of the hose, the hose will be angled to the local horizontal. When the end of hose couples with the receiver, refueling starts. Due to the huge strains of the hose while refueling, vibration of the hose can produce premature fatigue problems. Therefore, we propose boundary control to suppress the vibration of the hose described by PDEs. With the proposed control, the closed-loop stability is proved based on the Lyapunov direct method.

The rest of the chapter is organized as follows. The PDE dynamic model of a flexible aerial refueling hose is derived in Sect. 5.2. In Sect. 5.3, a boundary control scheme is designed and analyzed. Numerical simulations are demonstrated in Sect. 5.4 to show the effectiveness of the proposed controller.

5.2 Problem Formulation

A typical aerial refueling hose system in Fig. 5.1 is the connection between a tanker and a receiver aircraft. As shown in Fig. 5.1, $X_g - Y_g$ represents an inertial reference coordinate system. $X - Y$ represents the local coordinate system which is attached to and moves horizontally with the receiver aircraft, and $x - y$ the body-fixed coordinate

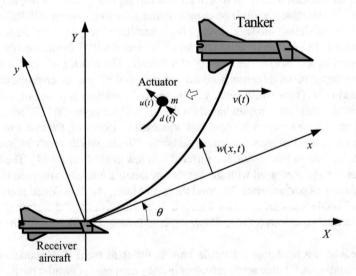

Fig. 5.1 Diagram of a flexible aerial refueling hose

system attached to point of junction between the hose and the receiver aircraft. In this chapter, we consider the transverse degree of freedom only. The orientation of the hose without deflections with respect to the local horizontal is denoted by θ. The control $u(t)$ is implemented from the actuator in the tanker, that is the top boundary of the hose. The tanker and the receiver aircraft have the same speed $v(t)$ and relative to $X_g - Y_g$. The receiver aircraft is at the position $r(t)$ relative to $X_g - Y_g$. $w(x, t)$ is the elastic deflection with respective to the frame $x - y$ at the position x for time t, and the position vector of the hose $p(x, t)$ respective to the frame $X - Y$ at the position x for time t is described by

$$p(x, t) = \begin{pmatrix} p_1(x, t) \\ p_2(x, t) \end{pmatrix} = \begin{pmatrix} x \cos \theta - w(x, t) \sin \theta \\ x \sin \theta + w(x, t) \cos \theta \end{pmatrix}$$

The absolute position vector of a point along the hose is denoted by

$$z(x, t) = \begin{pmatrix} z_1(x, t) \\ z_2(x, t) \end{pmatrix} = \begin{pmatrix} p_1(x, t) + r(t) \\ p_2(x, t) \end{pmatrix}$$

The kinetic energy of the hose system $E_k(t)$ can be represented as

$$E_k(t) = \frac{\rho}{2} \int_0^L \left(z_{1t}^2(x, t) + z_{2t}^2(x, t) \right) dx + \frac{1}{2} m \left(z_{1t}^2(L, t) + z_{2t}^2(L, t) \right) \tag{5.1}$$

The potential energy of the hose system $E_p(t)$ due to the axial force can be obtained from

$$E_p(t) = \frac{1}{2} \int_0^L P(x, t)[w_x(x, t)]^2 dx \tag{5.2}$$

where $P(x, t)$ is the tension of the hose that can be expressed as [11]

$$P(x, t) = f_t(x, t) + \rho x (g \sin \theta + \ddot{r}(t) \cos \theta) \tag{5.3}$$

where g is the acceleration of gravity, and ρ is the density of the hose. $f_t(x, t)$ is the skin friction drag in the tangential direction. The virtual work done on the system is given by

$$\delta W(t) = -\int_0^L f_n(x, t) \delta w(x, t) dx + (u(t) + d(t)) \delta w(L, t) \tag{5.4}$$

Then, the Hamilton's principle is applied as

$$\int_{t_1}^{t_2} (\delta E_k(t) - \delta E_p(t) + \delta W(t)) dt = 0 \tag{5.5}$$

We further obtain the following PDEs of the hose system as

$$\rho w_{tt}(x, t) = P_x(x, t)w_x(x, t) + P(x, t)w_{xx}(x, t) + Q(x, t) \tag{5.6}$$

$$Q(x, t) = -f_n(x, t) + \rho(g\cos\theta + \ddot{r}(t)\sin\theta) \tag{5.7}$$

and boundary conditions of the hose system as

$$mw_{tt}(L, t) - m\ddot{r}(t)\sin\theta + P(L, t)w_x(L, t) - u(t) - d(t) = 0 \tag{5.8}$$

$$w(0, t) = 0 \tag{5.9}$$

Remark 5.1 It is noted that, related to suppressing the transversal vibrations, only the variations of $w(x, t)$ is considered. The position of the aircraft $r(t)$ is prespecified function and, therefore, it is not necessary to consider its variation when evaluation (5.5), that is $\delta r(t) = \delta \dot{r}(t) = 0$.

Assumption 5.1 The disturbance $d(t)$ and its rate of change $\dot{d}(t)$ are bounded so that there exist positive constants $\bar{d}$ and $\bar{d}_v$ satisfying $|d(t)| \leq \bar{d}$ and $|\dot{d}(t)| \leq \bar{d}_v$, respectively.

Assumption 5.2 If the kinetic energy of the system (5.6)–(5.9), given by Eq. (5.1) is bounded $\forall(x, t) \in [0, L] \times [0, \infty)$, then $w_t(x, t)$ is bounded $\forall(x, t) \in [0, L] \times [0, \infty)$.

Assumption 5.3 If the potential energy of the system (5.6)–(5.9), given by Eq. (5.2) is bounded $\forall(x, t) \in [0, L] \times [0, \infty)$, then $w_x(x, t)$ is bounded $\forall(x, t) \in [0, L] \times [0, \infty)$.

Assumption 5.4 For the positive definite function $P(x, t)$, we assume that $P(x, t)$, $P_x(x, t)$ and $P_t(x, t)$ are bounded by known, constant lower, and upper bounds as follows,

$$0 \leq P_{\min} \leq P(x, t) \leq P_{\max}$$

$$0 \leq P_{x\,\min} \leq P_x(x, t) \leq P_{x\,\max}$$

$$0 \leq P_{t\,\min} \leq P_t(x, t) \leq P_{t\,\max}$$

$\forall(x, t) \in [0, L] \times [0, \infty)$.

Assumption 5.5 We assume that the function $Q(x, t)$ is bounded so that there exist a positive constant $Q_{\max}$ satisfying $|Q(x, t)| \leq Q_{\max}$, $\forall(x, t) \in [0, L] \times [0, \infty)$.

Remark 5.2 From the definition of $f_n(x, t)$ and $\dot{v}(t)$, we can obtain the value of the $Q(x, t)$ according to (5.7). If velocity $v(t)$ and acceleration $\dot{v}(t)$ of the tanker are known, then it is possible to calculate the maximum value of the $Q(x, t)$.

5.3 Boundary Control Design

The control objective is to propose a controller $u(t)$ to suppress the elastic vibration $w(x, t)$ of the flexible aerial refueling hose in the presence of the high speed of the aerial refueling hose system. In this section, we use the Lyapunov's direct method to design a boundary control law $u(t)$ on the top boundary of the hose and analyze the closed loop stability of the system.

First, we design a disturbance observer to estimate the disturbance. The estimates of $d(t)$ is denoted as $\hat{d}(t)$. The basic method to design observer is correcting the error between the estimated output and the actual output, so we design that

$$\dot{\hat{d}}(t) = K\left(d(t) - \hat{d}(t)\right) \tag{5.10}$$

where $K > 0$.

Then we define auxiliary vector as

$$\varphi(t) = \hat{d}(t) - Kmw_t(L, t) \tag{5.11}$$

With the system model (5.8), we can obtain

$$d(t) = mw_{tt}(L, t) - m\ddot{r}(t)\sin\theta + P(L, t)w_x(L, t) - u(t) \tag{5.12}$$

Differentiate (5.11) with respect to time, and substitute Eqs. (5.10) into it, we have

$$\dot{\varphi}(t) = \dot{\hat{d}}(t) - Kmw_{tt}(L, t)$$
$$= K\left(d(t) - \hat{d}(t)\right) - Kmw_{tt}(L, t) \tag{5.13}$$

Substitute Eqs. (5.12) into (5.13), we have

$$\dot{\varphi}(t) = \dot{\hat{d}}(t) - Kmw_{tt}(L, t)$$
$$= K\left(P(L, t)w_x(L, t) - m\ddot{r}(t)\sin\theta - u(t)\right) - K\hat{d}(t) \tag{5.14}$$

Then the disturbance observer dynamics is given as follows:

$$\dot{\varphi}(t) = K(P(L, t)w_x(L, t) - m\ddot{r}(t)sin\theta - u(t)) - K\hat{d}(t)$$
$$\hat{d}(t) = \varphi(t) + Kmw_t(L, t) \tag{5.15}$$

Consider the system dynamics described by (5.6)–(5.9) with Assumptions (1)–(5) and the disturbance observer (5.15), the boundary control law is designed as

$$u(t) = P(L, t)w_x(L, t) - m\ddot{r}(t)sin\theta - mw_{xt}(L, t) - ku_0 - \hat{d}(t) \tag{5.16}$$

where k is the control gain, and the auxiliary term is defined as

$$u_0 = w_t(L, t) + w_x(L, t) \tag{5.17}$$

Remark 5.3 All the signals in the boundary controller and the observer can be measured by sensors or obtained by a backward difference algorithm. $w(L, t)$ can be sensed by a laser displacement at the top boundary of the hose. $w_x(L, t)$ can be measured by an inclinometer at the top boundary of the hose. $w_{xt}(L, t)$ and $w_t(L, t)$ can be calculated with a backward difference algorithm. In practice, the effect of measurement noise from sensors is inevitable which will affect the control implementation.

Theorem 5.1 *Suppose the system (5.6)–(5.9) satisfies Assumptions (1)–(5). With the proposed boundary control law (5.16) and the disturbance observer (5.15), then the following properties should hold.*

(1) The closed-loop system is uniformly bounded, that is $w(x, t)$ remains in the compact set Ω_1 defined by

$$\Omega_1 = \{w(x, t) \in R : |w(x, t)| \leq C_1, \forall (x, t) \in [0, L] \times [0, \infty)\}$$

where the constant $C_1 = \sqrt{\frac{2L}{\beta P_{\min}\beta_2} \left(V(0)e^{-\lambda t} + \frac{\varepsilon}{\lambda}\right)}$, $V(0)$ is the initial value of the Lyapunov function $V(t)$.

(2) The closed-loop system is uniformly ultimate bounded, that is $w(x, t)$ eventually converges to the compact set Ω_2 defined by

$$\Omega_2 = \left\{w(x, t) \in R : \lim_{t \to \infty} |w(x, t)| \leq C_2, \forall x \in [0, L]\right\}$$

where the constant $C_2 = \sqrt{\frac{2L\varepsilon}{\beta P_{\min}\beta_2\lambda}}$, and the parameters β, β_2, λ and ε are defined in the following process of proof.

Proof The proof process can be fond in Appendix.

Remark 5.4 The proof process is the control law design process based on the Lyapunov's direct method. When the parameters α and β are chosen as proper values, from Eqs. (5.48) and (5.53), we can see that the increase of the control gain k will bring about a larger γ_3, which will result in a larger λ_1. Then the value of λ will increase, which will reduce the size of Ω_2, accelerate the rate of convergence and produce a better vibration reduction performance. Then we can conclude that the deflection of the hose $w(x, t)$ can be made arbitrarily small when the design control parameters are appropriately selected to satisfy the inequalities (5.43)–(5.47) in Appendix. However, a very large control gain k could lead to a high gain control problem. In practical applications, considering measurement noise and input constraints, we should choose the parameter k as small as possible on the condition that certain performance indicators can be satisfied.

Remark 5.5 The proposed boundary control law (5.16) has more benefits than the existing control schemes [5, 8]. The previous studies did not consider the acceleration $\dot{v}(t)$ of the hose system. In this chapter, the terms $P(L, t)w_x(L, t)$ and $w_{xt}(L, t)$ play an important role in the vibration suppressing, especially in the presence of the acceleration. And the control design is based on the PDEs, which avoids the spillover problems associated with traditional truncated model-based approaches caused by ignoring high-frequency models in controller and observer design.

Remark 5.6 The orientation of the hose with respect to the local horizontal θ depends on the speed of the hose before coupling. However, the orientation θ remains unchanged if the hose is towed horizontally at a constant speed during coupling.

5.4 Simulation

We use the finite difference method to simulate the system performance. By choosing the proper temporal and spatial step size to approximate the solution of the PDE system, the effectiveness of the proposed control law (5.16) with the disturbance observe (5.15) is demonstrated by the finite difference method. The disturbance $d(t)$ is given as $d(t) = 1 + 1\sin(0.1t) + 3\sin(0.3t)$. The initial conditions are given as $w(x, 0) = 1.025 \times 10^{-5}x^2$ and $\dot{w}(x, 0) = 0$. The parameters of the flexible hose are listed in Table 5.1.

For analyzing and verifying the control performance, the dynamic responses of the system are simulated in the following two cases:

Case 1: Consider the acceleration $\dot{v}(t)$

(1) With the PID control [5]

$$u(t) = -k_p w(L, t) - k_d \dot{w}(L, t) - k_f \int_0^t w(L, \tau)d\tau - \hat{d}(t) \tag{5.18}$$

Table 5.1 Parameters of a flexible aerial refueling hose

Parameter	Description	Value
L	The length of the hose	16 m
ρ	The mass of the unit length	5.2 kg/m
D	The diameter of the hose	0.067 m
m	The point mass of the actuator	16.8 kg
g	Acceleration of gravity	9.8 m/s²
ρ_{air}	The air density	1.29 kg/m³
C_f	The skin friction coefficient	0.005
C_d	The pressure drag coefficient	0.45

and the disturbance observer (5.15) with the parameters $k_p = 200$, $k_d = 200$, $k_f = 100$ and $K = 40$ at speed $v(t) = 100 + 4.7t$ during coupling.

(2) Applying the proposed control (5.16) and the disturbance observer (5.15) with the parameters $k = 200$ and $K = 40$ at speed $v(t) = 100 + 4.7t$ during coupling.

Case 2: Ignore the acceleration $\dot{v}(t)$

(1) With the PID control (5.18) and the disturbance observer (5.15) with the parameters $k_p = 200$, $k_d = 200$, $k_f = 100$ and $K = 40$ at speed $v(t) = 100$ during coupling.

(2) Applying the proposed control (5.16) and the disturbance observer (5.15) with the parameters $k = 200$ and $K = 40$ at speed $v(t) = 100$ during coupling.

(3) For analyzing the effect of the control gain k, the dynamic responses of the system are simulated in the following four cases: $k = 50$, $k = 100$, $k = 200$ and $k = 300$. Other parameters are the same.

The simulation results are shown in Figs. 5.2, 5.3, 5.4, 5.5, 5.6, 5.7, 5.8 and 5.9.

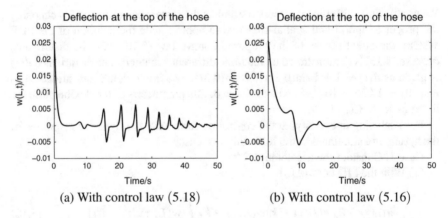

(a) With control law (5.18)   (b) With control law (5.16)

Fig. 5.2 Deflection at the top of the hose for case 1

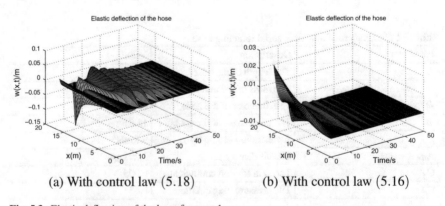

(a) With control law (5.18)   (b) With control law (5.16)

Fig. 5.3 Elastic deflection of the hose for case 1

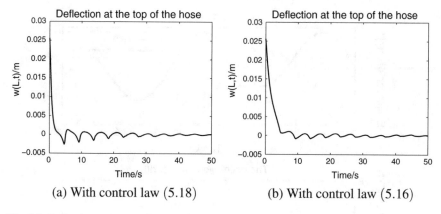

Fig. 5.4 Deflection at the top of the hose for case 2

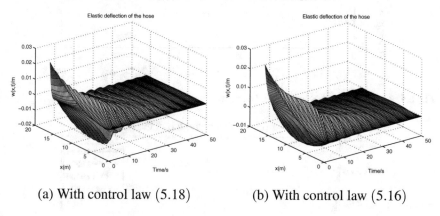

(a) With control law (5.18) (b) With control law (5.16)

Fig. 5.5 Elastic deflection of the hose for case 2

Figures 5.2 and 5.4 show deflection at the top of the hose. The elastic deflection of the hose $w(x, t)$ is shown in Figs. 5.3 and 5.5. From Figs. 5.2b and 5.3b, we can see that the proposed control scheme (5.16) can regulate the vibration greatly within 15 s, and $w(x, t)$ numerically converges to a small neighborhood of zero after 20 s, which means that the good performance of vibration suppressing can be obtained with the proposed control law. Figures 5.2a and 5.3a indicate that the PID control law makes the vibration $w(x, t)$ converge to a small neighborhood of zero after a much longer time. Moreover, Figs. 5.4 and 5.5 show that the both PID control law (5.18) and the proposed control law (5.16) can quickly suppress the vibration and keep the system vibration in a small neighborhood of zero when the acceleration is not considered. From above analysis we can conclude that the control scheme proposed in this paper has a better control performance especially in the presence of the acceleration.

And Fig. 5.6 shows the disturbance estimation and the estimation error, which indicates that the estimates of the disturbance converge to the true values. Using the

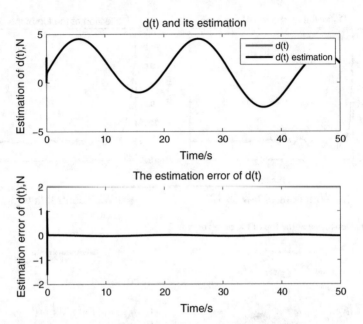

Fig. 5.6 The disturbance estimation and estimation error

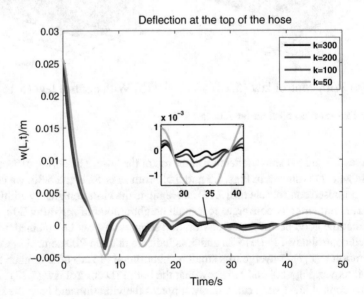

Fig. 5.7 Deflection at the top of the hose

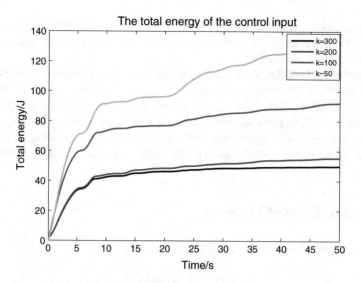

Fig. 5.8 The total energy of the control input

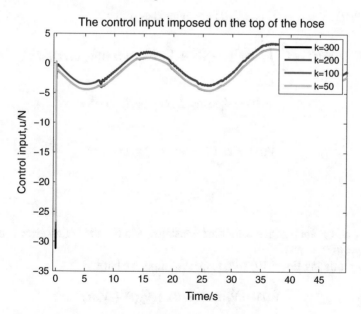

Fig. 5.9 The control input imposed on the top of the hose

disturbance observer rather than a robust control method to solve the problem of disturbance is advantageous to the safe use of the actuators and the attenuation of vibration of the controlled plant.

Form Figs. 5.7 and 5.8, we can see that as the control gain increases, a better control performance can be obtained. However, the control performance is not changed obviously when $k \geq 200$. Moreover, the actuator needs to provide a larger input which may result in input saturation when a very large control gain k is chosen. Therefore, in practical applications, we can choose the parameter k as small as possible on the condition that certain performance indicators can be satisfied.

Appendix: Proof of Theorem

Proof Consider the Lyapunov candidate function as

$$V(t) = V_1(t) + V_2(t) + V_3(t) + V_o(t) \tag{5.19}$$

where $V_1(t)$, $V_2(t)$, $V_3(t)$ and $V_o(t)$ are defined as follows

$$V_1(t) = \frac{\beta}{2} \int_0^L \rho w_t^2(x, t)dx + \frac{\beta}{2} \int_0^L P(x, t)[w_x(x, t)]^2 dx$$

$$V_2(t) = \frac{\beta}{2} m(w_t(L, t) + w_x(L, t))^2$$

$$V_3(t) = \alpha \int_0^L \rho x w_t(x, t) w_x(x, t)dx$$

$$V_o = \frac{1}{2} \tilde{d}^2(t)$$

where α and β are positive weighting constants, and the estimation error is defined as $\tilde{d}(t) = d(t) - \hat{d}(t)$.

Differentiating Eq. (5.19) with respect to time, we have

$$\dot{V}(t) = \dot{V}_1(t) + \dot{V}_2(t) + \dot{V}_3(t) + \dot{V}_o(t) \tag{5.20}$$

The term $\dot{V}_1(t)$ is rewritten as

$$\dot{V}_1(t) = \dot{V}_{11} + \dot{V}_{12} + \dot{V}_{13}$$

where

$$\dot{V}_{11} = \beta\rho \int_0^L w_t(x, t) w_{tt}(x, t)dx \tag{5.21}$$

$$\dot{V}_{12} = \beta \int_0^L P(x,t)w_x(x,t)\,w_{xt}(x,t)\,dx \qquad (5.22)$$

$$\dot{V}_{13} = \frac{\beta}{2} \int_0^L P_t(x,t)[w_x(x,t)]^2 dx \qquad (5.23)$$

Substituting Eq. (5.6) into (5.21), we get

$$\dot{V}_{11} = \beta\rho \int_0^L w_t(x,t)w_{tt}(x,t)dx$$
$$= \beta \int_0^L w_t(x,t)\,(P_x(x,t)w_x(x,t) + P(x,t)w_{xx}(x,t) + Q(x,t))\,dx$$

and integrating Eq. (5.22) by parts with the boundary conditions, we obtain

$$\dot{V}_{12} = \beta\,[P(L,t)w_x(L,t)\,w_t(L,t) - P(0,t)w_x(0,t)\,w_t(0,t)]$$
$$- \beta \int_0^L w_t(x,t)\,P_x(x,t)w_x(x,t)\,dx - \beta \int_0^L w_t(x,t)\,P(x,t)w_{xx}(x,t)\,dx$$

Then, we have

$$\dot{V}_1(t) = \beta \int_0^L w_t(x,t)Q(x,t)dx + \frac{\beta}{2}\int_0^L P_t(x,t)[w_x(x,t)]^2 dx$$
$$+ \frac{\beta P(L,t)}{2}\left[u_0^2 - (w_x(L,t))^2 - (w_t(L,t))^2\right] \qquad (5.24)$$

According to Lemma 2.4, we obtain

$$\dot{V}_1(t) \le \frac{\beta}{\sigma_1}\int_0^L w_t^2(x,t)dx + \beta\sigma_1\int_0^L Q^2(x,t)dx + \frac{\beta}{2}\int_0^L P_t(x,t)[w_x(x,t)]^2 dx$$
$$+ \frac{\beta P(L,t)}{2}\left[u_0^2 - (w_x(L,t))^2 - (w_t(L,t))^2\right] \qquad (5.25)$$

where σ_1 is a positive constant.

Considering the boundary condition (5.8), we have

$$\dot{V}_2(t) = \beta u_0\,(mw_{tt}(L,t) + mw_{xt}(L,t))$$
$$= \beta u_0\,(-P(L,t)w_x(L,t) + m\ddot{r}(t)\sin\theta)$$
$$+ \beta u_0\,(u(t) + d(t) + mw_{xt}(L,t)) \qquad (5.26)$$

Substituting (5.16) into (5.26), we obtain

$$\begin{aligned}
\dot{V}_2(t) &= -\beta k u_0^2 + \beta u_0 \tilde{d}(t) \\
&\le -\beta(k-1)u_0^2 + \beta \tilde{d}(t)^2
\end{aligned} \tag{5.27}$$

To go on, the term $\dot{V}_3(t)$ is rewritten as

$$\dot{V}_3(t) = \dot{V}_{31} + \dot{V}_{32} + \dot{V}_{33} + \dot{V}_{34} \tag{5.28}$$

where

$$\dot{V}_{31} = \alpha \int_0^L x w_x(x,t) P_x(x,t) w_x(x,t) \, dx \tag{5.29}$$

$$\dot{V}_{32} = \alpha \int_0^L x w_x(x,t) P(x,t) w_{xx}(x,t) \, dx \tag{5.30}$$

$$\dot{V}_{33} = \alpha \int_0^L x w_x(x,t) Q(x,t) dx \tag{5.31}$$

$$\dot{V}_{34} = \alpha \int_0^L \rho x w_t(x,t) w_{xt}(x,t) dx \tag{5.32}$$

Using the boundary conditions and integrating Eq. (5.30) by parts, we get

$$\begin{aligned}
\dot{V}_{32} = \alpha L[w_x(L,t)]^2 P(L,t) - \alpha \int_0^L w_x(x,t) P(x,t) w_x(x,t) \, dx \\
- \alpha \int_0^L x w_x(x,t) P_x(x,t) w_x(x,t) \, dx - \alpha \int_0^L x w_{xx}(x,t) P(x,t) w_x(x,t) \, dx
\end{aligned} \tag{5.33}$$

Combining (5.30) and (5.33), we have

$$\begin{aligned}
\dot{V}_{32} = \frac{\alpha L}{2} P(L,t)[w_x(L,t)]^2 - \frac{\alpha}{2} \int_0^L P(x,t)[w_x(x,t)]^2 dx \\
- \frac{\alpha}{2} \int_0^L x P_x(x,t)[w_x(x,t)]^2 dx
\end{aligned} \tag{5.34}$$

According to Lemma 2.4, we obtain

$$\dot{V}_{33} \le \frac{\alpha L}{\sigma_2} \int_0^L Q^2(x,t) dx + \alpha L \sigma_2 \int_0^L [w_x(x,t)]^2 dx \tag{5.35}$$

where σ_2 is a positive constant. Integrating (5.32) by parts, we obtain

$$\dot{V}_{34} = \alpha \rho L w_t^2(L, t) - \alpha \rho \int_0^L w_t^2(x, t)dx - \alpha \rho \int_0^L x w_t(x, t) w_{xt}(x, t)dx \quad (5.36)$$

Considering (5.32), we then get

$$\dot{V}_{34} = \frac{\alpha \rho}{2} L w_t^2(L, t) - \frac{\alpha \rho}{2} \int_0^L w_t^2(x, t)dx \quad (5.37)$$

Substituting (5.29), (5.34), (5.35) and (5.37) into (5.28), we obtain

$$\begin{aligned}
\dot{V}_3(t) \leq{}& \alpha \int_0^L x P_x(x, t)[w_x(x, t)]^2 dx + \frac{\alpha L}{2} P(L, t)[w_x(L, t)]^2 \\
&- \frac{\alpha}{2} \int_0^L P(x, t)[w_x(x, t)]^2 dx - \frac{\alpha}{2} \int_0^L x P_x(x, t)[w_x(x, t)]^2 dx \\
&+ \frac{\alpha L}{\sigma_2} \int_0^L Q^2(x, t)dx + \alpha L \sigma_2 \int_0^L [w_x(x, t)]^2 dx \\
&+ \frac{\alpha \rho}{2} L w_t^2(L, t) - \frac{\alpha \rho}{2} \int_0^L w_t^2(x, t)dx
\end{aligned} \quad (5.38)$$

Considering estimation error $\tilde{d}(t) = d(t) - \hat{d}(t)$, we then obtain

$$\begin{aligned}
\dot{V}_o(t) ={}& \tilde{d}(t)\dot{\tilde{d}}(t) = \tilde{d}(t)\dot{d}(t) - \tilde{d}(t)\dot{\hat{d}}(t) \\
={}& -\tilde{d}(\dot{\varphi}(t) + K m w_{tt}(L, t)) + \tilde{d}(t)\dot{d}(t) \\
={}& -K\tilde{d}(t)(P(L, t)w_x(L, t) - m\ddot{r}(t)\sin\theta - u(t)) + K\tilde{d}(t)\hat{d}(t) \\
&- K\tilde{d}(-P(L, t)w_x(L, t) + m\ddot{r}(t)\sin\theta + u(t) + d(t)) + \tilde{d}(t)\dot{d}(t) \\
={}& -K\tilde{d}(t)(d(t) - \hat{d}(t)) + \tilde{d}(t)\dot{d}(t) = -K\tilde{d}^2(t) + \tilde{d}(t)\dot{d}(t)
\end{aligned} \quad (5.39)$$

According to Lemma 2.4, we have

$$\begin{aligned}
\dot{V}_o(t) ={}& -K\tilde{d}^2(t) + \tilde{d}(t)\dot{d}(t) \\
\leq{}& -K\tilde{d}^2(t) + \sigma_3\tilde{d}^2(t) + \frac{1}{\sigma_3}\dot{d}^2(t) \\
\leq{}& -(K - \sigma_3)\tilde{d}^2(t) + \frac{1}{\sigma_3}\dot{d}_v^2(t)
\end{aligned} \quad (5.40)$$

Substituting (5.25), (5.26), (5.38) and (5.40) into (5.20), we obtain

$$
\begin{aligned}
\dot{V}(t) \leq &-\frac{1}{2} \int_0^L [\alpha P(x,t) - \beta \dot{P}(x,t) - \alpha x P_x(x,t) - 2\alpha L\sigma_2][w_x(x,t)]^2 dx \\
&- \left(\frac{\beta}{\sigma_1} + \frac{\alpha\rho}{2}\right) \int_0^L w_t^2(x,t)dx - \frac{\beta - \alpha L}{2} P(L,t)[w_x(L,t)]^2 \\
&- \left(\beta(k-1) - \frac{\beta P(L,t)}{2}\right) u_0^2 - \frac{1}{2}(\beta P(L,t) - \alpha\rho L) w_t^2(L,t) \\
&+ \left(\beta\sigma_1 + \frac{\alpha L}{\sigma_2}\right) \int_0^L Q^2(x,t)dx - (K - \sigma_3 - \beta)\tilde{d}^2(t) + \frac{1}{\sigma_3}\tilde{d}_v^2(t) \quad (5.41)
\end{aligned}
$$

To motivate the followings, we first focus our attention on the term $V_3(t)$. It satisfies the following inequality

$$
|V_3(t)| \leq \alpha\rho L \int_0^L w_t^2(x,t)dx + \alpha\rho L \int_0^L [w_x(x,t)]^2 dx \leq \beta_1 V_1(t)
$$

where $\beta_1 = \frac{2\alpha\rho L}{\beta \min(\rho, P_{\min})}$. We then obtain

$$
-\beta_1 V_1(t) \leq V_3(t) \leq \beta_1 V_1(t)
$$

Assuming that α is a small positive weighting constant satisfying $0 < \alpha < \frac{\beta \min(\rho, P_{\min})}{2\rho L}$, we can obtain $0 < \beta_1 < 1$, and

$$
\beta_2 (V_1(t) + V_2(t) + V_o(t)) \leq V(t) \leq \beta_3 (V_1(t) + V_2(t) + V_o(t)) \quad (5.42)
$$

where $\beta_2 = \min(1 - \beta_1, 1) = 1 - \beta_1$ and $\beta_3 = \max(1 + \beta_1, 1) = 1 + \beta_1$.

We design parameters α, β and k to satisfy the following inequality:

$$
\alpha P_{\min} - \beta \dot{P}_{\max} - \alpha L P'_{\max} - 2\alpha L\sigma_2 \geq \delta \quad (5.43)
$$

$\forall (x,t) \in [0, L] \times [0, \infty)$, for a positive constant δ, and the following conditions:

$$
\beta - \alpha L \geq 0 \quad (5.44)
$$

$$
\beta P_{\min} - \alpha\rho L \geq 0 \quad (5.45)
$$

$$
k - \frac{P_{\max}}{2} - 1 \geq 0 \quad (5.46)
$$

$$
K - \sigma_3 - \beta \geq 0 \quad (5.47)
$$

Equation (5.41) can be rewritten as

$$\dot{V}(t) \le -\gamma_1 \frac{\beta}{2} \int_0^L P(x,t)[w_x(x,t)]^2 dx - \gamma_2 \frac{\beta\rho}{2} \int_0^L w_t^2(x,t) dx$$
$$- \gamma_3 \frac{\beta}{2} m u_0^2 - \gamma_4 \frac{1}{2}\tilde{d}^2(t) + \varepsilon \tag{5.48}$$

where $\gamma_1 = \frac{\delta}{\beta P_{max}}$, $\gamma_2 = \left(\frac{2}{\sigma_1 \rho} + \frac{\alpha}{\beta}\right)$, $\gamma_3 = \frac{1}{m}(2\beta(k-1) - \beta P_{max})$, $\gamma_4 = 2$ $(K - \sigma_3 - \beta)$, and $\varepsilon = \left(\beta\sigma_1 + \frac{\alpha L}{\sigma_2}\right)L Q_{max}^2 + \frac{1}{\sigma_3}\bar{d}_v^2(t)$

We further obtain

$$\dot{V}(t) \le -\gamma_1 \frac{\beta}{2} \int_0^L P(x,t)[w_x(x,t)]^2 dx - \gamma_2 \frac{\beta\rho}{2} \int_0^L w_t^2(x,t) dx$$
$$- \gamma_3 \frac{\beta}{2} m u_0^2 + \varepsilon$$
$$\le -\lambda_1 [V_1(t) + V_2(t) + V_o(t)] + \varepsilon \tag{5.49}$$

where $\lambda_1 = \min(\gamma_1, \gamma_2, \gamma_3, \gamma_4)$.

Combining (5.42) and (5.49), we have

$$\dot{V}(t) \le -\lambda V(t) + \varepsilon \tag{5.50}$$

where $\lambda = \lambda_1/\beta_3 > 0$.

To go on, multiply Eq. (5.50) by $e^{\lambda t}$, we obtain

$$\frac{\partial}{\partial t}\left((V(t)e^{\lambda t})\right) \le \varepsilon e^{\lambda t} \tag{5.51}$$

Integrating of the inequality (5.51), we have

$$V(t) \le \left(V(0) - \frac{\varepsilon}{\lambda}\right)e^{-\lambda t} + \frac{\varepsilon}{\lambda} \le V(0)e^{-\lambda t} + \frac{\varepsilon}{\lambda} \in \mathscr{L}_\infty$$

This implies that $V(t)$ is bounded.

According to Lemma 2.5, we have

$$\frac{\beta P_{min}}{2L}w^2(x,t) \le \frac{\beta}{2}\int_0^L P(x,t)[w_x(x,t)]^2 dx \le V_1(t)$$
$$\le V_1(t) + V_2(t) + V_o(t) \le \frac{V(t)}{\beta_2} \in \mathscr{L}_\infty \tag{5.52}$$

It follows that, $|w(x,t)| \le \sqrt{\frac{2L}{\beta P_{min}\beta_2}\left(V(0)e^{-\lambda t} + \frac{\varepsilon}{\lambda}\right)}$, $\forall(x,t) \in [0,L] \times [0,\infty)$, so $w(x,t)$ is uniformly bounded. From (5.52), we further obtain

$$\lim_{t \to \infty} |w(x, t)| \le \sqrt{\frac{2L\varepsilon}{\beta P_{min}\beta_2\lambda}}, \forall(x, t) \in [0, L] \times [0, \infty) \tag{5.53}$$

This completes the proof.

References

1. Fravolini ML, Ficola A, Campa G, Napolitano MR, Seanor B (2004) Modeling and control issues for autonomous aerial refueling for UAVs using a probe–drogue refueling system. Aerosp Sci Technol 8(7):611–618
2. He W, Zhang S, Ge SS (2014) Adaptive control of a flexible crane system with the boundary output constraint. IEEE Trans Ind Electron 61(8):4126–4133
3. Kriel SC, Engelbrecht JAA, Jones T (2013) Receptacle normal position control for automated aerial refueling. Aerosp Sci Technol 29(1):296–304
4. Meirovitch L, Baruh H (1983) On the problem of observation spillover in self-adjoint distributed-parameter systems. J Optim Theory Appl 39(2):269–291
5. Ro K, Kamman JW (2010) Modeling and simulation of hose-paradrogue aerial refueling systems. J Guid, Control, Dyn 33(1):53–63
6. Ro K, Kuk T, Kamman JW (2010) Active control of aerial refueling hose-drogue systems. In: AIAA guidance, navigation, and control conference. Toronto, Ontario Canada, p 8400
7. Ro K, Kuk T, Kamman JW (2011) Dynamics and control of hose-drogue refueling systems during coupling. J Guid, Control, Dyn 34(6):1694–1708
8. Wang H, Dong X, Xue J, Liu J (2014) Dynamic modeling of a hose-drogue aerial refueling system and integral sliding mode backstepping control for the hose whipping phenomenon. Chin J Aeronaut 27(4):930–946
9. Williams P, Sgarioto D, Trivailo P (2006) Optimal control of an aircraft-towed flexible cable system. J Guid, Control, Dyn 29(2):401–410
10. Williams P, Trivailo P (2007) Dynamics of circularly towed aerial cable systems, part i: optimal configurations and their stability. J Guid, Control, Dyn 30(3):753–765
11. Zhu WD, Ni J, Huang J (2001) Active control of translating media with arbitrarily varying length. J Vib Acoust 123(3):347–358
12. Zhu ZH, Meguid SA (2007) Modeling and simulation of aerial refueling by finite element method. Int J Solids Struct 44(24):8057–8073

Chapter 6
Vibration Control of a Flexible Aerial Refueling Hose with Input Saturation

6.1 Introduction

In Chap. 5, we developed a PDE model for a flexible aerial refueling hose system during coupling. As we know, a hose-drogue aerial refueling system consists of a flexible hose and an active drogue control actuator, which are the most universal refueling equipments of probe-drogue refueling (PDR). The probe and drogue systems are comparatively simpler and more compact than the flying boom, and their arrangement on the tanker enables multiple aircraft to be refueled simultaneously. The significant drawback is that PDR requires a skillful piloting technique of maneuvering a probe into the center of a moving drogue with an acceptable closure rate. However, due to the flexible property of the hose, the deflection of the flexible hose has a significant influence on the dynamics and control performance of the AAR, which brings difficulties to the coupling. In order to solve this problem, we will establish a model for a flexible aerial refueling hose system before coupling, and develop an efficient boundary control scheme to suppress vibrations. Moreover, considering that common active drogue control actuators which consists of a set of aerodynamic control surface [5, 7] or other kinds of controllable drogues [6] cannot provide enough control input, which degrades the performance of the control system or leads to the instability.

In this chapter, we investigate the boundary control problem for a flexible aerial refueling hose with input saturation.

The model of a flexible hose is different from the existing models such as an Euler–Bernoulli beam [4], flexible wings of a robotic aircraft [2] or a gantry crane [1], which is affected by the horizontal velocity and gravity. Different from the axially moving string in [3], the hose moves horizontally at an angle, which makes the control problem in this work more difficult to handle compared to the previous works. Then a novel boundary controller is proposed for the flexible hose based on PDEs. With the proposed control, the closed-loop stability is proved based on the Lyapunv's direct method and the deflection eventually converges to an arbitrarily small neighborhood

© Tsinghua University Press 2020
Z. Liu and J. Liu, *PDE Modeling and Boundary Control for Flexible Mechanical System*, Springer Tracts in Mechanical Engineering, https://doi.org/10.1007/978-981-15-2596-4_6

around the origin. The main contributions of this chapter are that: (1) boundary control with a smooth hyperbolic function is designed to suppress the vibration of the flexible hose based on the original PDE model, and (2) an auxiliary system with a Nussbaum function is used to compensate for the nonlinear term arising from the input saturation.

The rest of the chapter is organized as follows. The PDE dynamic model of a flexible aerial refueling hose is derived in Sect. 6.2. In Sect. 6.3, a boundary control scheme is designed and analyzed. Numerical simulations are demonstrated in Sect. 6.4 to show the effectiveness of the proposed controller.

6.2 Problem Formulation

As shown in Fig. 6.1, $X_g - Y_g$ represents an inertial reference coordinate system. $X - Y$ represents the local coordinate system which is attached to and moves with the tanker, and $x - y$ the body-fixed coordinate system attached to the hose. In this study, we consider the transverse degree of freedom only. The orientation of the hose with respect to the local horizontal is denoted by θ. The control $u(t)$ is implemented by a drogue control actuator, $w(x, t)$ is the elastic deflection of the hose with respect to the frame $x - y$ at the position x for time t. The tanker has the speed $v(t)$ relative to $X_g - Y_g$ and its position vector relative to the frame $X_g - Y_g$ is $\left(r(t), h_0 \right)^T$. The

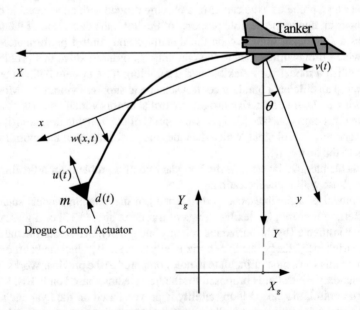

Fig. 6.1 Diagram of a flexible aerial refueling hose

position vector of the hose $p(x, t)$ respective to the frame $X - Y$ at the position x for time t is described by

$$p(x, t) = \begin{pmatrix} p_1(x, t) \\ p_2(x, t) \end{pmatrix} = \begin{pmatrix} x \cos \theta - w(x, t) \sin \theta \\ x \sin \theta + w(x, t) \cos \theta \end{pmatrix}$$

The absolute position vector of a point along the hose respective to the frame $X_g - Y_g$ is denoted by

$$z(x, t) = \begin{pmatrix} z_1(x, t) \\ z_2(x, t) \end{pmatrix} = \begin{pmatrix} r(t) - p_1(x, t) \\ h_0 - p_2(x, t) \end{pmatrix}$$

The kinetic energy of the hose system $E_k(t)$ can be represented as

$$E_k(t) = \frac{\rho}{2} \int_0^L \left(z_{1t}^2(x, t) + z_{2t}^2(x, t) \right) dx + \frac{1}{2} m \left(z_{1t}^2(L, t) + z_{2t}^2(L, t) \right) \quad (6.1)$$

where ρ is the density of the hose, L is the length of the hose, and m is the point mass of the drogue.

The gravitational potential energy of the hose system $E_{pg}(t)$ can be expressed as

$$E_{pg}(t) = \rho g \int_0^L z_2(x, t) dx + m g z_2(L, t) \quad (6.2)$$

and the potential energy of the hose system $E_{pf}(t)$ due to the axial force can be obtained from

$$E_{pf}(t) = \frac{1}{2} \int_0^L P(x, t)[w_x(x, t)]^2 dx \quad (6.3)$$

where $P(x, t)$ is the tension of the hose that can be expressed as

$$P(x, t) = (m + \rho (L - x)) (g \sin \theta - \ddot{r}(t) \cos \theta) \\ + f_{drog}(t) \cos \theta + f_t(x, t) \quad (6.4)$$

where g is the acceleration of gravity, $f_t(x, t)$ is the skin friction drag in the tangential direction, and $f_{drog}(t)$ is the drag of the drogue. So the potential energy of the hose system is

$$E_p(t) = E_{pg}(t) + E_{pf}(t) \quad (6.5)$$

The virtual work done on the system is given by

$$\delta W(t) = -\int_0^L f_n(x, t) \delta w(x, t) dx + \left(u(t) + d(t) - f_{drog}(t) \sin \theta \right) \delta w(L, t) \quad (6.6)$$

where $f_n(x, t)$ is the pressure drag in the normal direction.

Then, the Hamilton's principle is applied as

$$\int_{t_1}^{t_2} (\delta E_k(t) - \delta E_p(t) + \delta W(t))dt = 0 \tag{6.7}$$

where $\delta (\cdot)$ represents the variation of $(\cdot)$. We further obtain the following PDEs of the hose system as

$$\rho w_{tt}(x, t) = P_x(x, t)w_x (x, t) + P(x, t)w_{xx} (x, t) + Q(x, t) \tag{6.8}$$

$$Q(x, t) = - f_n(x, t) + \rho (g \cos \theta - \ddot{r}(t) \sin \theta) \tag{6.9}$$

and boundary conditions of the hose system as

$$mw_{tt}(L, t) + m\ddot{r}(t) \sin \theta + P(L, t)w_x (L, t) - mg \cos \theta$$
$$= - f_{drog}(t) \sin \theta + u(t) + d(t) \tag{6.10}$$

$$w (0, t) = 0 \tag{6.11}$$

In this chapter, we consider a flexible hose system with the input saturation, where the input saturation model is described as follows

$$u(t) = u_g(u_0(t)) = u_M \times \tanh \left(\frac{u_0(t)}{u_M} \right) \tag{6.12}$$

where u_M is a known bound of $u(t)$, and $u_0(t)$ is the designed control command.

Remark 6.1 It is noted that, related to suppressing the transversal vibrations, only the variations of $w(x, t)$ is considered. The position of the aircraft $r(t)$ is prespecified function, therefore, it is not necessary to consider its variation when evaluation (6.7), that is $\delta r (t) = \delta \dot{r}(t) = 0$. Moreover, the flight altitude h_0 is considered as a constant, therefore, $\dot{h}_0 = 0$.

Assumption 6.1 The disturbance $d(t)$ is bounded so that there exists a positive constant $\bar{d}$ satisfying $|d(t)| \leq \bar{d}$.

Assumption 6.2 For the positive definite function $P(x, t)$, we assume that $P(x, t)$, $P_x(x, t)$ and $P_t(x, t)$ are bounded by known, constant lower, and upper bounds as follows,

$$0 \leq P_{\min} \leq P(x, t) \leq P_{\max}$$

$$0 \leq P_{x \min} \leq P_x(x, t) \leq P_{x \max}$$

$$0 \leq P_{t \min} \leq P_t(x, t) \leq P_{t \max}$$

$\forall (x, t) \in [0, L] \times [0, \infty)$.

Assumption 6.3 We assume that the function $Q(x, t)$ is bounded so that there exist a positive constant Q_{max} satisfying $|Q(x, t)| \leq Q_{max}$, $\forall(x, t) \in [0, L] \times [0, \infty)$.

Remark 6.2 From the definition of $f_n(x, t)$ and $\dot{v}(t)$, we can obtain the value of $Q(x, t)$ according to (6.9). If velocity $v(t)$ and acceleration $\dot{v}(t)$ of the tanker are known, then it is possible to calculate the maximum value of the $Q(x, t)$.

6.3 Control Design

The control objective is to propose a controller $u(t)$ to suppress the elastic vibration $w(x, t)$ of the flexible aerial refueling hose in the presence of the high speed of the aerial refueling hose system and input saturation. In this section, we use the backstepping method to design a boundary control law $u(t)$ on the top boundary of the hose and use the Lyapunov's direct method to analyze the closed-loop stability of the system.

As the usual backstepping approach, the following transform of coordinate is made:

$$
\begin{aligned}
z_1 &= x_1 = w(L, t) \\
z_2 &= x_2 - \tau_1 = w_t(L, t) - \tau_1 \\
z_3 &= u_g(u_0(t)) - \tau_2
\end{aligned}
\tag{6.13}
$$

where τ_1 and τ_2 are the virtual control.

Step 1. We choose the virtual control law τ_1 as

$$
\tau_1 = -c_1 z_1 - w_x(L, t)
\tag{6.14}
$$

where $c_1 > 0$.

Then we consider a Lyapunov function candidate as

$$
V_{b1} = \frac{1}{2} z_1^2
\tag{6.15}
$$

The derivative of (6.15) is

$$
\dot{V}_{b1} = z_1 \dot{z}_1 = z_1 x_2 = z_1 (z_2 + \tau_1)
\tag{6.16}
$$

Substituting (6.14) into (6.16), we have

$$
\dot{V}_1 = -c_1 z_1^2 - z_1 w_x(L, t) + z_1 z_2
\tag{6.17}
$$

Step 2. Similarly, we choose the virtual control law τ_2 as

$$\tau_2 = -(c_2 + l) z_2 - z_1 + \beta c_1 P(L, t) z_1 + P(L, t) w_x (L, t)$$
$$+ m\ddot{r}(t) \sin\theta - mg \cos\theta + f_{drog}(t) \sin\theta + m\dot{\tau}_1 \tag{6.18}$$

where $c_2 > 0$ and $l > 0$. Then we consider a Lyapunov function candidate as

$$V_{b2} = V_{b1} + \frac{1}{2} m z_2^2 \tag{6.19}$$

Noting that $z_3 = u_g(u_0) - \tau_2$, the derivative of (6.19) is

$$\dot{V}_{b2} = \dot{V}_{b1} + m z_2 \dot{z}_2$$
$$= -c_1 z_1^2 - z_1 w_x(L, t) + z_1 z_2 + z_2 (m\dot{x}_2 - m\dot{\tau}_1)$$
$$= -c_1 z_1^2 - z_1 w_x(L, t) + z_1 z_2$$
$$+ z_2 (-P(L, t) w_x (L, t) - m\ddot{r}(t) \sin\theta + mg \cos\theta)$$
$$+ z_2 (-f_{drog}(t) \sin\theta + z_3 + \tau_2 + d(t) - m\dot{\tau}_1) \tag{6.20}$$

Substituting (6.18) into (6.20), we obtain

$$\dot{V}_{b2} = -c_1 z_1^2 - (c_2 + l) z_2^2 - z_1 w_x(L, t) + z_2 z_3 + z_2 d(t)$$
$$+ \beta c_1 P(L, t) z_1 x_2 + \beta c_1 P(L, t) c_1 z_1^2$$
$$+ \beta c_1 P(L, t) z_1 w_x(L, t) \tag{6.21}$$

Using the inequality $z_2 d(t) \le l z_2^2 + \frac{1}{4l} d^2(t)$, (6.21) can be rewritten as

$$\dot{V}_{b2} \le -c_1 z_1^2 - c_2 z_2^2 - z_1 w_x(L, t) + z_2 z_3 + \frac{1}{4l} d^2(t)$$
$$+ \beta c_1 P(L, t) z_1 x_2 + \beta c_1 P(L, t) c_1 z_1^2$$
$$+ \beta c_1 P(L, t) z_1 w_x(L, t) \tag{6.22}$$

We design an auxiliary system as

$$\dot{u}_0(t) = -c u_0(t) + \omega \tag{6.23}$$

where $c > 0$.

Considering (6.13), (6.18) and (6.23), we obtain

$$m\dot{z}_3 = m \frac{\partial u_g}{\partial u_0(t)} (-c u_0(t) + \omega) - m\dot{\tau}_2$$

$$= m\xi \left(-cu_0(t) + \omega\right) - \frac{\partial \tau_2}{\partial x_2} d(t) - \vartheta \tag{6.24}$$

$$\vartheta = \frac{\partial \tau_2}{\partial x_2} \left(u_g \left(u_0(t)\right) - P(L, t) w_x \left(L, t\right)\right) + m\ddot{r}(t)\sin\theta$$

$$+ m \frac{\partial \tau_2}{\partial P(L, t)} P_t(L, t) + m \frac{\partial \tau_2}{\partial x_1} x_2 + m \frac{\partial \tau_2}{\partial w_x \left(L, t\right)} w_{xt} \left(L, t\right)$$

$$+ m \frac{\partial \tau_2}{\partial w_{xt} \left(L, t\right)} w_{xtt} \left(L, t\right) + m \frac{\partial \tau_2}{\partial f_{drog}(t)} \dot{f}_{drog}(t)$$

$$\xi = \frac{\partial u_g \left(u_0(t)\right)}{\partial u_0(t)} = \frac{4}{\left(e^{u_0(t)/u_M} + e^{-u_0(t)/u_M}\right)^2} > 0$$

Then we use a Nussbaum function $N(\chi)$ to design the control law ω in (6.23), and ω is designed as

$$\omega = N(\chi)\bar{\omega} \tag{6.25}$$

$$\bar{\omega} = \xi cu_0 + \frac{1}{m}\vartheta - \frac{1}{m}c_3 z_3 - \frac{1}{m}z_2 - \frac{1}{m}l\left(\frac{\partial \tau_2}{\partial x_2}\right)^2 z_3 \tag{6.26}$$

where $c > 0$ and $N(\chi)$ is a Nussbaum function defined as

$$N(\chi) = \chi^2 \cos(\chi) \tag{6.27}$$

$$\dot{\chi} = \gamma_\chi m z_3 \bar{\omega}$$

where γ_χ is a positive real design parameter. And the Nussbaum function satisfies the two properties

$$\lim_{k \to \pm\infty} \sup \frac{1}{k} \int_0^k N(s)\, ds = \infty$$

$$\lim_{k \to \pm\infty} \inf \frac{1}{k} \int_0^k N(s)\, ds = -\infty$$

Step 3. We choose a Lyapunov function candidate as

$$V_b = V_{b2} + \frac{1}{2}mz_3^2 \tag{6.28}$$

The derivative of (6.28) is

$$\dot{V}_b \leq -c_1 z_1^2 - c_2 z_2^2 - z_1 w_x(L, t) + z_2 z_3 + \frac{1}{4l}d^2(t) + mz_3\dot{z}_3$$

$$+ \beta c_1 P(L, t)z_1 x_2 + \beta c_1 P(L, t)c_1 z_1^2 + \beta c_1 P(L, t)z_1 w_x(L, t)$$

$$= -c_1 z_1^2 - c_2 z_2^2 - z_1 w_x(L, t) + \frac{1}{4l}d^2(t) + z_2 z_3$$

$$+ z_3 (m\dot{z}_3 + m\bar{\omega}) - mz_3\bar{\omega} + \beta c_1 P(L, t)z_1 x_2$$
$$+ \beta c_1 P(L, t)c_1 z_1^2 + \beta c_1 P(L, t)z_1 w_x(L, t) \tag{6.29}$$

Substituting (6.24) and (6.26) into (6.29), and using (6.25), we have

$$\dot{V}_b \leq -c_1 z_1^2 - c_2 z_2^2 - c_3 z_3^2 - z_1 w_x(L, t) + \frac{1}{4l}d^2(t)$$
$$+ \beta c_1 P(L, t)z_1 x_2 + \beta c_1 P(L, t)c_1 z_1^2 + \beta c_1 P(L, t)z_1 w_x(L, t)$$
$$- l\left(\frac{\partial \tau_2}{\partial x_2}\right)^2 z_3^2 + z_3 m (\xi N(\chi) - 1)\bar{\omega} - z_3 \frac{\partial \tau_2}{\partial x_2} d(t)$$

Considering the two inequalities

$$-d(t)\frac{\partial \tau_2}{\partial x_2}z_3 \leq \frac{1}{4l}d^2(t) + l\left(\frac{\partial \tau_2}{\partial x_2}\right)^2 z_3^2$$
$$z_1 w_x(L, t) \leq \frac{1}{\sigma_3}z_1^2 + \sigma_3[w_x(L, t)]^2$$

and noting that $\dot{\chi} = \gamma_\chi mz_3\bar{\omega}$, we obtain

$$\dot{V}_b \leq -c_1 z_1^2 - c_2 z_2^2 - c_3 z_3^2 + \beta c_1 P(L, t)c_1 z_1^2$$
$$+ \beta c_1 P(L, t)z_1 x_2 + \beta c_1 P(L, t)z_1 w_x(L, t) + \frac{1}{\sigma_3}z_1^2$$
$$+ \sigma_3[w_x(L, t)]^2 + \frac{1}{\gamma_\chi}(\xi N(\chi) - 1)\dot{\chi} + \frac{1}{2l}\bar{d}^2 \tag{6.30}$$

Now we define the Lyapunov candidate function:

$$V(t) = V_1(t) + V_2(t) + V_b(t) \tag{6.31}$$

where $V_1(t)$ and $V_2(t)$ are defined as follows

$$V_1(t) = \frac{\beta}{2}\int_0^L \rho w_t^2(x, t)dx + \frac{\beta}{2}\int_0^L P(x, t)[w_x(x, t)]^2 dx$$
$$V_2(t) = \alpha \int_0^L \rho(x - L) w_t(x, t)w_x(x, t)dx$$

where α and β are positive weighting constants.

Lemma 6.1 *The time derivative of the Lyapunov function defined in (6.31) is upper bounded with*

$$V(t) \leq V(0)e^{-\lambda t} + \frac{\varepsilon_0}{\lambda} \tag{6.32}$$

where $\lambda > 0$, *and* $\varepsilon_0 > 0$, *which are defined in the process of proof.*

Proof The proof of the Lemma 6.1 can be found in the Appendix 1.

Theorem 6.1 *Suppose the system (6.8)–(6.11) satisfies Assumptions 6.1–6.3. With the proposed boundary control law (6.12), (6.14), (6.18), (6.23) and (6.26), the closed-loop system is globally stable and the following properties hold.*

(1) The control input is bounded, and its bound is described as:

$$|u(t)| = \left| u_g\left(u_0(t)\right) \right| = u_M \left| \tanh\left(\frac{u_0(t)}{u_M}\right) \right| \leq u_M$$

(2) The closed-loop system is uniformly ultimately bounded, that is $w(x, t)$ *eventually converges to the positive constant C, that is*

$$\lim_{t \to \infty} \sup_{x \in [0, L]} |w(x, t)| \leq C$$

where the constant $C = \sqrt{\frac{2L\varepsilon}{\beta P_{\min}\beta_2\lambda}}$, *and the parameters* β, β_2, λ *and* ε *are defined in the process of proving Lemma 6.1.*

Proof The proof of the Theorem can be found in the Appendix 2.

Remark 6.3 From the proof process, it is shown that the increase of the control gains c_2, c_3 and l will reduce the size of C and produce a better vibration reduction performance if the parameters c_1, α and β are chosen as proper values. Then we can conclude that the deflection of the hose $w(x, t)$ can be made arbitrarily small when the design control parameters are appropriately selected. However, very large control gains c_2, c_3 and l could lead to a high gain control problem. In practical applications, we should choose the parameters carefully to achieve the suitable control performance.

6.4 Simulation

We use the finite difference method to simulate the system performance. By choosing the proper temporal and spatial step size to approximate the solution of the PDE system, the effectiveness of the proposed control law (6.12) is demonstrated by the finite difference method. The disturbance $d(t)$ is given as $d(t) = 0.1 * sin(t)$. The initial conditions are given as $y(x, 0) = 4 \times 10^{-3}x^2$ and $y_t(x, 0) = 0$. The restriction on the input u is given by $|u| \leq u_M = 5$. The parameters of the flexible hose are listed in Table 6.1.

Table 6.1 Parameters of a flexible aerial refueling hose

Parameter	Description	Value
L	The length of the hose	16 m
ρ	The mass of the unit length	5.2 kg/m
m	The mass of the drogue	39.5 kg
g	Acceleration of gravity	9.8 m/s^2
ρ_{air}	The air density	0.909 kg/m^3
D	The diameter of the hose	0.066 m
D_{drog}	The diameter of the drogue	0.61 m
C_f	The skin friction coefficient	0.005
C_d	The pressure drag coefficient	0.45
C_{drog}	The drag coefficient	0.43

For analyzing and verifying the control performance, the dynamic responses of the system are simulated in the following two cases:

Case 1: Without control input: $u(t) = 0$,

Case 2:

(1) With the PID control

$$u(t) = -k_p w(L, t) - k_d w_t(L, t) - k_f \int_0^t w(L, \tau) d\tau \qquad (6.33)$$

with the parameters $k_p = 10$, $k_d = 10$ and $k_f = 10$ at speed $v(t) = 100$ m/s. And the restriction on the value of the input with the PID control is also given by $|u| \leq u_M = 5$.

(2) Applying the proposed control (6.12) with the parameters $c = 2$, $c_1 = 15$, $c_2 = 15$, $c_3 = 15$ and $l = 45$ at speed $v(t) = 100$ m/s.

The dynamic responses without input control are shown in the Fig. 6.2. It is clear that the vibration of the hose is large.

The simulation results of case 2 are shown in Figs. 6.3, 6.4 and 6.5.

Figure 6.3 shows deflection at the top of the hose. The elastic deflection of the hose $w(x, t)$ is shown in Fig. 6.4. From Figs. 6.3a and 6.4a, we can see that the proposed control scheme (6.12) can regulate the vibration greatly within 10 s, and $w(x, t)$ numerically converges to a small neighborhood of zero after 20 s, which means that the good performance of vibration suppressing can be obtained with the proposed control law. Figures 6.3b and 6.4b indicate that the PID control law under the saturation limit $|u| \leq u_M = 5$ makes the vibration $w(x, t)$ converge to a small neighborhood of zero after a much longer time. Moreover, Fig. 6.5 shows the control input imposed on the top of the hose. From above analysis we can conclude that the effectiveness of the control scheme proposed in this chapter can be guaranteed in handling input saturation.

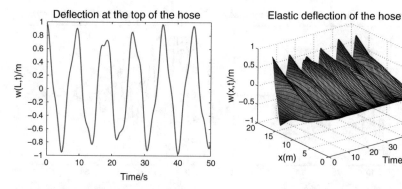

(a) Deflection at the top of the hose   (b) Elastic deflection of the hose

Fig. 6.2 The dynamic responses without control input

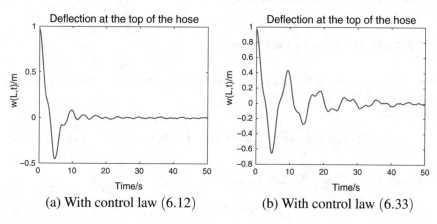

(a) With control law (6.12)   (b) With control law (6.33)

Fig. 6.3 Deflection at the top of the hose

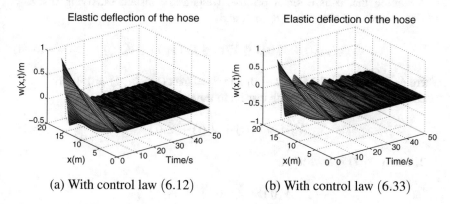

(a) With control law (6.12)   (b) With control law (6.33)

Fig. 6.4 Elastic deflection of the hose

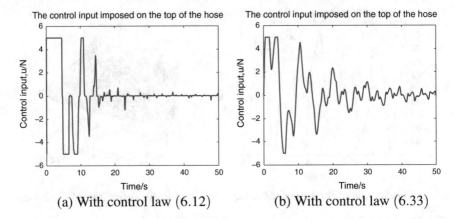

(a) With control law (6.12)   (b) With control law (6.33)

Fig. 6.5 The control input imposed on the top of the hose

Appendix 1: Proof of Lemma 6.1

Proof To motivate the followings, we first focus our attention on the term $V_2(t)$. It satisfies the following inequality

$$|V_2(t)| \leq \alpha \rho L \int_0^L w_t^2(x, t)dx + \alpha \rho L \int_0^L [w_x(x, t)]^2 dx \leq \beta_1 V_1(t)$$

where $\beta_1 = \frac{2\alpha \rho L}{\beta \min(\rho, P_{\min})}$. We then obtain

$$-\beta_1 V_1(t) \leq V_2(t) \leq \beta_1 V_1(t)$$

Assuming that α is a small positive weighting constant satisfying $0 < \alpha < \frac{\beta \min(\rho, P_{\min})}{2\rho L}$, we can obtain $0 < \beta_1 < 1$, and

$$\beta_2 (V_1(t) + V_b(t)) \leq V(t) \leq \beta_3 (V_1(t) + V_b(t)) \tag{6.34}$$

where $\beta_2 = \min (1 - \beta_1, 1) = 1 - \beta_1$ and $\beta_3 = \max (1 + \beta_1, 1) = 1 + \beta_1$. Differentiating Eq. (6.31) with respect to time, we have

$$\dot{V}(t) = \dot{V}_1(t) + \dot{V}_2(t) + \dot{V}_b(t) \tag{6.35}$$

The term $\dot{V}_1(t)$ is rewritten as

$$\dot{V}_1(t) = \dot{V}_{11} + \dot{V}_{12} + \dot{V}_{13}$$

where

$$\dot{V}_{11} = \beta\rho \int_0^L w_t(x, t)w_{tt}(x, t)dx \tag{6.36}$$

$$\dot{V}_{12} = \beta \int_0^L P(x, t)w_x(x, t) w_{xt}(x, t) dx \tag{6.37}$$

$$\dot{V}_{13} = \frac{\beta}{2} \int_0^L P_t(x, t)[w_x(x, t)]^2 dx \tag{6.38}$$

Substituting Eq. (6.8) into (6.36), we get

$$\dot{V}_{11} = \beta\rho \int_0^L w_t(x, t)w_{tt}(x, t)dx$$
$$= \beta \int_0^L w_t(x, t) \left(P_x(x, t)w_t(x, t) + P(x, t)w_{xx}(x, t) + Q(x, t)\right) dx$$

and integrating Eq. (6.37) by parts with the boundary conditions, we obtain

$$\dot{V}_{12} = \beta \left[P(L, t)w_x(L, t) w_t(L, t) - P(0, t)w_x(0, t) w_t(0, t)\right]$$
$$- \beta \int_0^L w_t(x, t) P_x(x, t)w_x(x, t) dx$$
$$- \beta \int_0^L w_t(x, t) P(x, t)w_{xx}(x, t) dx$$

Then, we have

$$\dot{V}_1(t) = \beta \int_0^L w_t(x, t)Q(x, t)dx + \frac{\beta}{2} \int_0^L P_t(x, t)[w_x(x, t)]^2 dx$$
$$+ \frac{\beta P(L, t)}{2} \left[z_2^2 - (w_x(L, t))^2 - (w_t(L, t))^2 - c_1^2 w^2(L, t)\right]$$
$$- \beta c_1 P(L, t)w_t(L, t) w(L, t) - \beta c_1 P(L, t)w_x(L, t) w(L, t) \tag{6.39}$$

According to Lemma 2.4, we obtain

$$\dot{V}_1(t) \le \frac{\beta}{\sigma_1} \int_0^L w_t^2(x, t)dx + \beta\sigma_1 \int_0^L Q^2(x, t)dx$$
$$+ \frac{\beta}{2} \int_0^L P_t(x, t)[w_x(x, t)]^2 dx - \beta c_1 P(L, t)w_t(L, t) w(L, t)$$
$$+ \frac{\beta P(L, t)}{2} \left[z_2^2 - (w_x(L, t))^2 - (w_t(L, t))^2 - c_1^2 w^2(L, t)\right]$$
$$- \beta c_1 P(L, t)w_x(L, t) w(L, t) \tag{6.40}$$

where σ_1 is a positive constant.

To go on, the term $\dot{V}_2(t)$ is rewritten as

$$\dot{V}_2(t) = \dot{V}_{21} + \dot{V}_{22} + \dot{V}_{23} + \dot{V}_{24} \tag{6.41}$$

where

$$\dot{V}_{21} = \alpha \int_0^L (x - L) w_x(x, t) P_x(x, t) w_x(x, t)\, dx \tag{6.42}$$

$$\dot{V}_{22} = \alpha \int_0^L (x - L) w_x(x, t) P(x, t) w_{xx}(x, t)\, dx \tag{6.43}$$

$$\dot{V}_{23} = \alpha \int_0^L (x - L) w_x(x, t) Q(x, t)\, dx \tag{6.44}$$

$$\dot{V}_{24} = \alpha \int_0^L \rho\, (x - L) w_t(x, t) w_{xt}(x, t)\, dx \tag{6.45}$$

Using the boundary conditions and integrating Eq. (6.43) by parts, we get

$$
\begin{aligned}
\dot{V}_{22} = &-\alpha \int_0^L w_x(x, t) P(x, t) w_x(x, t)\, dx \\
&-\alpha \int_0^L (x - L) w_x(x, t) P_x(x, t) w_x(x, t)\, dx \\
&-\alpha \int_0^L (x - L) w_{xx}(x, t) P(x, t) w_x(x, t)\, dx
\end{aligned}
\tag{6.46}
$$

Combining (6.43) and (6.46), we have

$$\dot{V}_{22} = -\frac{\alpha}{2} \int_0^L P(x, t)[w_x(x, t)]^2 dx \tag{6.47}$$

$$\qquad\qquad -\frac{\alpha}{2} \int_0^L (x - L)\, P_x(x, t)[w_x(x, t)]^2 dx \tag{6.48}$$

According to Lemma 2.4, we obtain

$$\dot{V}_{23} \le \frac{\alpha L}{\sigma_2} \int_0^L Q^2(x, t)dx + \alpha L \sigma_2 \int_0^L [w_x(x, t)]^2 dx \tag{6.49}$$

where σ_2 is a positive constant. Integrating (6.45) by parts, we obtain

$$\dot{V}_{24} = -\alpha \rho \int_0^L w_t^2(x, t)dx - \alpha \rho \int_0^L (x - L) w_t(x, t) w_{xt}(x, t)dx$$

Considering (6.45), we then get

$$\dot{V}_{24} = -\frac{\alpha\rho}{2} \int_0^L w_t^2(x, t)dx \tag{6.50}$$

Substituting (6.42), (6.48), (6.49) and (6.50) into (6.41), we obtain

$$\begin{aligned}
\dot{V}_2(t) \leq{} & \alpha \int_0^L (x - L) P_x(x, t)[w_x (x, t)]^2 dx \\
& - \frac{\alpha}{2} \int_0^L P(x, t)[w_x (x, t)]^2 dx + \alpha L\sigma_2 \int_0^L [w_x(x, t)]^2 dx \\
& - \frac{\alpha}{2} \int_0^L (x - L) P_x(x, t)[w_x (x, t)]^2 dx \\
& + \frac{\alpha L}{\sigma_2} \int_0^L Q^2(x, t)dx - \frac{\alpha\rho}{2} \int_0^L w_t^2(x, t)dx
\end{aligned} \tag{6.51}$$

Substituting (6.30), (6.40) and (6.51) into (6.35), we obtain

$$\begin{aligned}
\dot{V} ={} & \dot{V}_1 + \dot{V}_2 + \dot{V}_b \\
\leq{} & -\frac{1}{2} \int_0^L [\alpha P(x, t) - \alpha(x - L)P_x(x, t)] [w_x (x, t)]^2 dx \\
& - \frac{1}{2} \int_0^L [-2\alpha L\sigma_2 - \beta P_t(x, t)] [w_x (x, t)]^2 dx \\
& - \left(\frac{\alpha\rho}{2} - \frac{\beta}{\sigma_1}\right) \int_0^L w_t^2(x, t)dx \\
& - \left(c_1 + \frac{\beta P(L, t)c_1^2}{2} - \beta P(L, t)c_1^2 - \frac{1}{\sigma_3}\right) z_1^2 \\
& - \left(c_2 - \frac{\beta P(L, t)}{2}\right) z_2^2 - c_3 z_3^2 - \beta P(L, t)\frac{w_t^2(L, t)}{2} \\
& - \left(\frac{\beta P(L, t)}{2} - \sigma_3\right) [w_x(L, t)]^2 \\
& + \frac{1}{\gamma_x} (\xi N (\chi) - 1) \dot{\chi} + \frac{1}{2l}\bar{d}^2 \\
& + \left(\frac{\alpha L}{\sigma_2} + \sigma_1\beta\right) \int_0^L Q^2(x, t)dx
\end{aligned} \tag{6.52}$$

We design parameters α and β to satisfy the following inequality:

$$\alpha P_{\min} - \beta P_{t\max} - \alpha L P_{x\max} - 2\alpha L\sigma_2 \geq \delta$$

$\forall (x, t) \in [0, L] \times [0, \infty)$, for a positive constant δ, and the following conditions:

$$\frac{\alpha \rho}{2} - \frac{\beta}{\sigma_1} > 0$$

$$c_1 + \frac{\beta P_{\min} c_1^2}{2} - \beta P_{\max} c_1^2 - \frac{1}{\sigma_3} > 0$$

$$c_2 - \frac{\beta P_{\max}}{2} \geq 0$$

$$\frac{\beta P_{\min}}{2} - \sigma_3 \geq 0$$

Equation (6.52) can be rewritten as

$$\dot{V}(t) \leq -\gamma_1 \frac{\beta}{2} \int_0^L P(x, t)[w_x(x, t)]^2 dx - \gamma_2 \frac{\beta \rho}{2} \int_0^L w_t^2(x, t) dx$$
$$- \gamma_3 \frac{1}{2} z_1^2 - \gamma_4 \frac{m}{2} z_2^2 - \gamma_5 \frac{m}{2} z_3^2$$
$$+ \frac{1}{\gamma_\chi} (\xi N(\chi) - 1) \dot{\chi} + \varepsilon \qquad (6.53)$$

where

$$\gamma_1 = \frac{\delta}{\beta P_{\max}}$$

$$\gamma_2 = \left(\frac{\alpha}{\beta} - \frac{2}{\sigma_1 \rho} \right)$$

$$\gamma_3 = 2 \left(c_1 + \frac{\beta P_{\min} c_1^2}{2} - \beta P_{\max} c_1^2 - \frac{1}{\sigma_3} \right)$$

$$\gamma_4 = \frac{2}{m} \left(c_2 - \frac{\beta P_{\max}}{2} \right)$$

$$\gamma_5 = \frac{2}{m} c_3$$

$$\varepsilon = \frac{1}{2l} \bar{d}^2 + \left(\frac{\alpha L}{\sigma_2} + \sigma_1 \beta \right) L Q_{\max}^2$$

We further obtain

$$\dot{V}(t) \leq -\lambda_1 [V_1(t) + V_b(t)] + \frac{1}{\gamma_\chi} (\xi N(\chi) - 1) \dot{\chi} + \varepsilon \qquad (6.54)$$

where $\lambda_1 = \min(\gamma_1, \gamma_2, \gamma_3, \gamma_4, \gamma_5)$.

Combining (6.34) and (6.54), we have

$$\dot{V}(t) \leq -\lambda V(t) + \frac{1}{\gamma_\chi} \left(\xi N (\chi) - 1 \right) \dot{\chi} + \varepsilon \tag{6.55}$$

where $\lambda = \lambda_1/\beta_3 > 0$.

Then multiplying Eq. (6.55) by $e^{\lambda t}$, we obtain

$$\frac{\partial}{\partial t} \left(\left(V(t) e^{\lambda t} \right) \right) \leq \varepsilon e^{\lambda t} + \frac{1}{\gamma_\chi} \left(\xi N (\chi) - 1 \right) \dot{\chi} e^{\lambda t} \tag{6.56}$$

Integrating of the inequality (6.56), we have

$$\begin{aligned} V(t) &\leq V(0) e^{-\lambda t} + \frac{\varepsilon}{\lambda} \left(1 - e^{-\lambda t} \right) \\ &\quad + \frac{e^{-\lambda t}}{\gamma_\chi} \int_0^t \left(\xi N (\chi) - 1 \right) \dot{\chi} e^{\lambda \tau} d\tau \\ &\leq V(0) e^{-\lambda t} + \frac{\varepsilon_0}{\lambda} \end{aligned} \tag{6.57}$$

where $\varepsilon_0 = \varepsilon + \frac{\lambda}{\gamma_\chi} \int_0^t \left(\xi N (\chi) - 1 \right) \dot{\chi} e^{-\lambda(t-\tau)} d\tau$.

Applying Lemma 2.8, we can conclude that $V(t)$, χ and $\int_0^t \left(\xi N (\chi) - 1 \right) \dot{\chi} d\tau$ are bounded on $[0, t)$.

This completes the proof.

Appendix 2: Proof of Theorem 6.1

Proof According to Lemma 6.1, we can conclude that z_1, z_2, z_3, $w(x, t)$, $w_t(x, t)$ and $w_x(x, t)$ are all bounded.

Note that

$$\left| u_g (u_0) \right| = u_M \left| \tanh \left(\frac{u_0}{u_M} \right) \right| \leq u_M \tag{6.58}$$

$$\left| \frac{\partial u_g (u_0)}{\partial u_0} \right| = \left| \frac{4}{\left(e^{u_0/u_M} + e^{-u_0/u_M} \right)^2} \right| \leq 1 \tag{6.59}$$

$$\left| \frac{\partial u_g (u_0)}{\partial u_0} u_0 \right| = \left| \frac{4 u_0}{\left(e^{u_0/u_M} + e^{-u_0/u_M} \right)^2} \right| \leq \frac{u_M}{2} \tag{6.60}$$

Then we can obtain that $\bar{\omega}$ is bounded from (6.58)–(6.60) and (6.26). This further implies that ω and $u_0(t)$ are bounded.

Combining with (6.32) and according to Lemma 2.5, we have

$$\frac{\beta P_{\min}}{2L} w^2(x, t) \leq \frac{\beta}{2} \int_0^L P(x, t)[w_x(x, t)]^2 dx \leq V_1(t)$$

$$\leq V_1(t) + V_b(t) \leq \frac{V(t)}{\beta_2} \tag{6.61}$$

Then we get

$$w(x, t) \leq \sqrt{\frac{2l(t)}{\beta P_{\min} \beta_2} \left(V(0)e^{-\lambda t} + \frac{\varepsilon_0}{\lambda} \right)} \tag{6.62}$$

It follows that, $\lim\limits_{t \to \infty} |w(x, t)| \leq \sqrt{\frac{2L\varepsilon_0}{\beta P_{\min} \beta_2 \lambda}}$, $\forall (x, t) \in [0, L] \times [0, \infty)$, so $w(x, t)$ is uniformly ultimate bounded.

This completes the proof.

References

1. He W, Ge SS (2016) Cooperative control of a nonuniform gantry crane with constrained tension. Automatica 66:146–154
2. He W, Zhang S (2016) Control design for nonlinear flexible wings of a robotic aircraft. IEEE Trans Control Syst Technol 25(1):351–357
3. He W, Zhang S, Ge SS (2014) Adaptive control of a flexible crane system with the boundary output constraint. IEEE Trans Ind Electron 61(8):4126–4133
4. Jin F-F, Guo B-Z (2015) Lyapunov approach to output feedback stabilization for the Euler-Bernoulli beam equation with boundary input disturbance. Automatica 52:95–102
5. Ro K, Kuk T, Kamman JW (2010) Active control of aerial refueling hose-drogue systems. In: AIAA Guidance, Navigation, and Control Conference, Toronto, Ontario, Canada, p 8400
6. Thomas PR, Bhandari U, Bullock S, Richardson TS, Du Bois JL (2014) Advances in air to air refuelling. Prog Aerosp Sci 71:14–35
7. Williamson WR, Reed E, Glenn GJ, Stecko SM, Musgrave J, Takacs JM (2010) Controllable drogue for automated aerial refueling. J Aircr 47(2):515–527

Chapter 7
Deadzone Compensation Based Boundary Control for Flexible Aerial Refueling Hose with Output Constraint

7.1 Introduction

In this chapter, we consider the vibration suppression problem of a flexible hose system subject to input deadzone, output constraint and external disturbances. Until now, many researchers have paid attention to control design for the system with input deadzone [7, 10]. In [10], neural networks are used to compensate the input deadzone. A method based on barrier Lyapunov function (BLF) is proposed in [8] to deal with output constraints. The control design and stability analysis in the papers mentioned above are based on ODEs, so the methods can not be directly used to PDEs. Moreover, there is little information about how to handle the input deadzone and output constraints simultaneously for flexible mechanical systems based on PDEs. In [3], boundary control is designed for a flexible robotic manipulator subject to input deadzone. But output constraint is not taken into account in this paper. Conversely, in [1, 2, 4, 5], boundary control for flexible mechanical systems with output constraints are designed based on the BLF, which do not consider input deadzone. Few published papers have discussed control design for flexible mechanical systems subject to input deadzone and output constraints based on PDEs. It is still a challenging research topic.

The current work investigates the vibration problem for a flexible hose system under input deadzone, output constraint and external disturbances. The flexible hose is described by PDEs. Firstly, a basic boundary control scheme, based on the back-stepping method, is proposed to suppress the hose's vibration in the presence of input deadzone. A radial basis function (RBF) neural network is used to handle the effect of the input deadzone. Then, a BLF is used to avoid exceeding output constraint. With the designed control, the proof of the stability is obtained based on the Lyapunov direct method. The main contributions of this chapter are: (1) boundary control scheme with a RBF neural network is designed to stabilize the flexible hose under input deadzone based on the original PDE model; (2) Using a BLF, all signals and states of the system are ensured with boundness, and output constraint violation is avoided.

© Tsinghua University Press 2020
Z. Liu and J. Liu, *PDE Modeling and Boundary Control for Flexible Mechanical System*, Springer Tracts in Mechanical Engineering,
https://doi.org/10.1007/978-981-15-2596-4_7

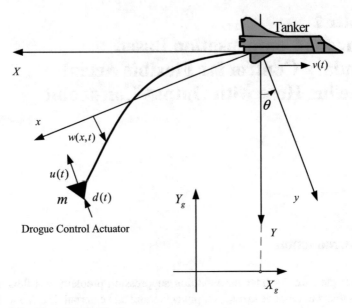

Fig. 7.1 A flexible aerial refueling hose system

The rest of this chapter is organized as follows. Section 7.2 introduces the dynamical model. Section 7.3 elucidates the design of a basic boundary control scheme. In Sect. 7.4, we present a new controller based on BLF to prevent constraint violation. Numerical simulation is given in Sect. 7.5.

7.2 Problem Formulation

In this chapter, we consider a flexible hose system in Chap. 6, as shown in Fig. 7.1, where the governing equation is given as

$$\rho w_{tt}(x, t) = P_x(x, t)w_x(x, t) + P(x, t)w_{xx}(x, t) + Q(x, t) \qquad (7.1)$$

where

$$P(x, t) = (m + \rho (L - x)) (g \sin \theta - \ddot{r}(t) \cos \theta)$$
$$+ f_{drog}(t) \cos \theta + f_t(x, t) \qquad (7.2)$$

$$Q(x, t) = -f_n(x, t) + \rho (g \cos \theta - \ddot{r}(t) \sin \theta) \qquad (7.3)$$

and boundary conditions as

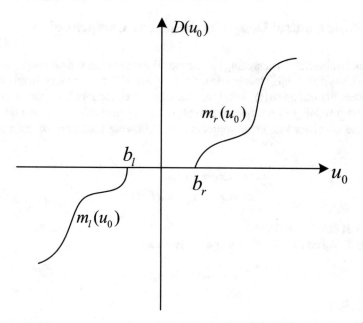

Fig. 7.2 Input deadzone model

$$mw_{tt}(L, t) + m\ddot{r}(t) \sin \theta + P(L, t)w_x(L, t) - mg \cos \theta$$
$$= -f_{drog}(t) \sin \theta + u(t) + d(t) \tag{7.4}$$

$$w(0, t) = w_{tt}(0, t) = 0 \tag{7.5}$$

where $f_t(x, t)$ and $f_n(x, t)$ are the skin friction drag and the pressure drag in the tangential direction and normal direction, respectively, and $f_{drog}(t)$ is the drag of the drogue.

Consider the hose system subject to the input deadzone, where the deadzone nonlinearity, as shown in Fig. 7.2, is described as follows [9]

$$u(t) = D(u_0) = \begin{cases} m_r(u_0 - b_r), & u_0 \geq b_r \\ 0, & b_l < u_0 < b_r \\ m_l(u_0 - b_l), & u_0 \leq b_l \end{cases} \tag{7.6}$$

where $u_0(t)$ is the control signal we will design, b_r and b_l are unknown parameters of the deadzone, $m_r(\cdot)$ and $m_l(\cdot)$ are unknown functions of the deadzone.

Assumption 7.1 ([9]) The deadzone parameters b_r and b_l satisfy the condition that $b_r > 0$ and $b_l < 0$.

7.3 Basic Control Design for Deadzone Compensation

The control objective of designing basic controller is to suppress displacement $w(x, t)$ of the hose system subject to input deadzone and external disturbances. In this section, a boundary control law $u(t)$ is designed based on the backstepping approach and the closed loop stability of the system is analyzed by Lyapunov's direct method.

As the usual backstepping approach, the following transform of coordinate is made:

$$z_1 = x_1 = w(L, t)$$
$$z_2 = x_2 - \tau = w_t(L, t) - \tau$$

where τ is the virtual control.

Step 1. We choose the virtual control law τ as

$$\tau = -c_1 z_1 - w_x(L, t) \tag{7.7}$$

where $c_1 > 0$.

Then a Lyapunov function candidate is chosen as

$$V_{b1} = \frac{1}{2} z_1^2 \tag{7.8}$$

The derivative of (7.8) is

$$\dot{V}_{b1} = z_1 \dot{z}_1 = z_1 x_2 = z_1 (z_2 + \tau) \tag{7.9}$$

Substituting (7.7) into (8.21), we have

$$\dot{V}_{b1} = -c_1 z_1^2 - z_1 w_x(L, t) + z_1 z_2 \tag{7.10}$$

Step 2. Let $D(u_0) = u_0 - \Delta u_0$, Δu_0 is the error. Choosing a positive definite function

$$V_{b2} = V_{b1} + \frac{1}{2} m z_2^2 \tag{7.11}$$

its time derivative is

$$
\begin{aligned}
\dot{V}_{b2} &= \dot{V}_{b1} + m z_2 \dot{z}_2 \\
&= -c_1 z_1^2 - z_1 w_x(L, t) + z_1 z_2 + z_2 (m \dot{x}_2 - m \dot{\tau}) \\
&= -c_1 z_1^2 - z_1 w_x(L, t) + z_1 z_2 \\
&\quad + z_2 \left(-P(L, t) w_x(L, t) - m \ddot{r}(t) \sin\theta + mg \cos\theta\right) \\
&\quad + z_2 \left(-f_{drog}(t) \sin\theta + u_0 - \Delta u_0 + d(t) - m \dot{\tau}\right)
\end{aligned} \tag{7.12}
$$

We choose the control law u_0 as

$$u_0 = -(c_2 + l) z_2 - z_1 + \beta c_1 P(L, t) z_1 - mg \cos \theta + f_{drog}(t) \sin \theta$$
$$+ P(L, t) w_x (L, t) + m\ddot{r}(t) \sin \theta + m\dot{\tau} + \Delta u_0 - d(t) \qquad (7.13)$$

where $c_2 > 0$ and $l > 0$.

Substituting (7.13) into (7.12), we obtain

$$\dot{V}_{b2} = -c_1 z_1^2 - (c_2 + l) z_2^2 - z_1 w_x (L, t)$$
$$+ \beta c_1 P(L, t) z_1 z_2 \qquad (7.14)$$

However, the terms of Δu_0 and $d(t)$ are unknown. We use a RBF neural network to approximate Δu_0, and Δu_0 can be represented by RBF with constant ideal weight W^{*T} and a basis function $\mathbf{h}(\chi)$, that is

$$\Delta u_0 = W^{*T} \mathbf{h}(\chi) + \varepsilon_u \qquad (7.15)$$

where ε_u is the RBF approximation error satisfying $|\varepsilon_u| \leq \varepsilon_{uM}$ with constant ε_{uM}, and $\mathbf{h}(\chi)$ is given by

$$\mathbf{h}(\chi) = \left[h_j(\chi)\right]^T \qquad (7.16)$$

$$h_j(\chi) = \exp\left(\frac{\|\chi - \mu_j\|^2}{\eta_j^2}\right), j = 1, 2, \ldots, n \qquad (7.17)$$

where μ_j represents the center of the jth basis function, η_j represents the spread of the basis function (e.g., Gaussian function), and $\chi = \left[z_1, z_2, \dot{\tau}, w'(L, t)\right]^T$.

To estimate the external boundary disturbance, we design a disturbance observer

$$\dot{\varphi}(t) = K(P(L, t) w_x(L, t) + m\ddot{r}(t) \sin\theta - u(t))$$
$$+ K(-mg \cos \theta + f_{drog}(t) \sin \theta) - K\hat{d}(t)$$
$$\hat{d}(t) = \varphi(t) + Kmw_t(L, t) \qquad (7.18)$$

where $K > 0$ and $\hat{d}(t)$ is the estimate of $d(t)$.

Step 3. Now, we propose a controller as

$$u_0 = -(c_2 + l) z_2 - z_1 + \beta c_1 P(L, t) z_1 - mg \cos \theta + f_{drog}(t) \sin \theta$$
$$+ P(L, t) w_x (L, t) + m\ddot{r}(t) \sin \theta + m\dot{\tau} - \hat{d}(t) + \hat{W}^T \mathbf{h}(\chi) \qquad (7.19)$$

and the network updating law is designed as

$$\dot{\hat{W}} = -\gamma_h \mathbf{h}(\chi) z_2 - \gamma_h \sigma_1 \hat{W} \qquad (7.20)$$

where $\hat{W}$ is the estimate of the ideal weight W^*, $\gamma_h > 0$ and $\sigma_1 > 0$.

Substituting (7.19) and (7.15) into (7.12) yields

$$\dot{V}_{b2} = -c_1 z_1^2 - (c_2 + l) z_2^2 - z_1 w_x(L, t) + \beta c_1 P(L, t) z_1 z_2$$
$$+ z_2 \tilde{d}(t) + z_2 \tilde{W}^T \mathbf{h}(\chi) - z_2 \varepsilon_u \tag{7.21}$$

Then we consider a positive definite function

$$V_b = V_{b2} + \frac{1}{2\gamma_h} \tilde{W}^T \tilde{W} + \frac{1}{2} \tilde{d}^2(t) \tag{7.22}$$

where $\tilde{W} = \hat{W} - W^*$ and $\tilde{d}(t) = d(t) - \hat{d}(t)$.

The derivative of V_b is

$$\dot{V}_b = -c_1 z_1^2 - (c_2 + l) z_2^2 - z_1 w_x(L, t) + \beta c_1 P(L, t) z_1 z_2$$
$$+ z_2 \tilde{d}(t) + z_2 \tilde{W}^T \mathbf{h}(\chi) - z_2 \varepsilon_u + \frac{1}{\gamma_h} \tilde{W}^T \dot{\hat{W}} + \tilde{d}(t) \dot{\tilde{d}}(t) \tag{7.23}$$

Using (7.4) and (7.18), we have

$$\dot{\tilde{d}}(t) = \dot{d}(t) - \dot{\hat{d}}(t) = \dot{d}(t) - (\dot{\varphi}(t) + K m w_{tt}(L, t))$$
$$= \dot{d}(t) - K(d(t) - \hat{d}(t)) = \dot{d}(t) - K \tilde{d}(t) \tag{7.24}$$

Substituting (7.20) and (7.24) into (7.23) yields

$$\dot{V}_b \leq -c_1 z_1^2 - (c_2 - \sigma_2) z_2^2 + \frac{1}{\sigma_3} z_1^2 + \sigma_3 [w_x(L, t)]^2$$
$$+ \beta c_1 P(L, t) z_1 z_2 + \frac{1}{\sigma_2} \varepsilon_u^2 - \left(K - \sigma_6 - \frac{1}{4l} \right) \tilde{d}^2(t)$$
$$- \frac{1}{2} \sigma_1 \tilde{W}^T \tilde{W} + \frac{1}{2} \sigma_1 W^{*T} W^* + \frac{1}{\sigma_6} \bar{d}_v^2 \tag{7.25}$$

where σ_2, σ_3 and σ_6 are positive constants, $\bar{d}$ and $\bar{d}_v$ are the maximum values of $d(t)$ and $\dot{d}(t)$, respectively.

Theorem 7.1 *Assume that the system (7.1)–(7.5) satisfies Assumption 7.1. Using the basic boundary control law (7.19), RBF neural network updating law (7.20), and the observer (7.18), the closed loop system is globally stable and the following properties hold.*

(1) The uniform boundness (UB) of the system is proven, that is $w(x, t)$ satisfies $|w(x, t)| \leq C_1$, $\forall (x, t) \in [0, L] \times [0, \infty)$, where the constant $C_1 = \sqrt{\frac{2L}{\beta P_{min} \beta_2}} \left(V(0) + \frac{\varepsilon}{\lambda} \right)$.

(2) The uniform ultimate boundedness (UUB) of the system is also proven, that is $w(x, t)$ satisfies $\lim\limits_{t \to \infty} |w(x, t)| \leq C_2, \forall x \in [0, L]$, where C_2, and the parameters β, β_2, λ and ε are defined in Appendix 1.

Proof See Appendix 1.

Remark 7.1 In the basic control design, the transient performance of the deflection $w(L, t)$ cannot be guaranteed. Therefore, we will employ a BLF to prevent the boundary output from violating the constraint.

7.4 Barrier Lyapunov Function (BLF) Based Control Design

The objectives of this section are to design boundary controller to (1) suppress the displacement $w(x, t)$ of the flexible hose, and (2) make the end-point displacement $w(L, t)$ satisfy the constraint $|w(L, t)| \leq b$ subject to input deadzone and external disturbance, where b is a positive constant.

Compared with the basic design, the major differences lie in the first two steps in the backstepping procedure. Thus the redesigned first two steps are as follows:

Step 1. Considering a barrier Lyapunov function as

$$\bar{V}_{b1} = \frac{1}{2} \ln \frac{b^2}{b^2 - z_1^2}$$

its time derivative is

$$\dot{\bar{V}}_{b1} = \frac{z_1 \dot{z}_1}{b^2 - z_1^2} = \frac{z_1 x_2}{b^2 - z_1^2} = \frac{z_1}{b^2 - z_1^2} (z_2 + \tau) \tag{7.26}$$

We choose the virtual control law τ as

$$\tau = -c_1 z_1 - (b^2 - z_1^2) w_x(L, t) \tag{7.27}$$

Substituting (7.27) into (7.26), we have

$$\dot{\bar{V}}_{b1} = -\frac{c_1 z_1^2}{b^2 - z_1^2} - z_1 w_x(L, t) + \frac{z_1 z_2}{b^2 - z_1^2} \tag{7.28}$$

Step 2. Then a Lyapunov function candidate is chosen as

$$\bar{V}_{b2} = \bar{V}_{b1} + \frac{1}{2} m z_2^2 \tag{7.29}$$

The derivative of (7.29) is

$$\dot{\bar{V}}_{b2} = -\frac{c_1 z_1^2}{b^2 - z_1^2} - z_1 w_x(L, t) + \frac{z_1 z_2}{b^2 - z_1^2}$$
$$+ z_2 \left(-P(L, t) w_x(L, t) - m\ddot{r}(t) sin\theta + mg\cos\theta \right)$$
$$+ z_2 \left(-f_{drog}(t) \sin\theta + u_0 - \Delta u_0 + d(t) - m\ddot{t} \right) \qquad (7.30)$$

Similarly, we choose the law u_0 as

$$u_0 = -(c_2 + l)z_2 - \frac{\beta P(L, t)z_2}{2(b^2 - z_1^2)} - \frac{z_1}{b^2 - z_1^2} + \frac{\beta c_1 P(L, t)z_1}{b^2 - z_1^2}$$
$$- mg\cos\theta + f_{drog}(t)\sin\theta + P(L, t)w_x(L, t)$$
$$+ m\ddot{r}(t)\sin\theta + m\ddot{t} - \hat{d}(t) + \hat{W}^T \mathbf{h}(\chi) \qquad (7.31)$$

Using (7.31) and (7.15) yields

$$\dot{\bar{V}}_{b2} \leq -\frac{c_1 z_1^2}{b^2 - z_1^2} - (c_2 + l)z_2^2 - z_1 w_x(L, t) - \frac{\beta P(L, t)z_2^2}{2(b^2 - z_1^2)}$$
$$+ \frac{\beta c_1 P(L, t)z_1 z_2}{b^2 - z_1^2} + z_2 \tilde{d}(t) + z_2 \tilde{W}^T \mathbf{h}(\chi) - z_2 \varepsilon_u \qquad (7.32)$$

Then a Lyapunov function candidate is chosen as

$$\bar{V}_b = \bar{V}_{b2} + \frac{1}{2\gamma_h} \tilde{W}^T \tilde{W} + \frac{1}{2} \tilde{d}^2(t) \qquad (7.33)$$

In the same way as the step 3 in the basic design, we have

$$\dot{\bar{V}}_b \leq -\frac{c_1 z_1^2}{b^2 - z_1^2} - (c_2 - \sigma_2) z_2^2 + \frac{z_1^2}{\sigma_3(b^2 - z_1^2)}$$
$$+ \sigma_3(b^2 - z_1^2)[w_x(L, t)]^2 - \frac{\beta P(L, t)z_2^2}{2(b^2 - z_1^2)}$$
$$+ \frac{\beta c_1 P(L, t)z_1 z_2}{b^2 - z_1^2} - \left(K - \sigma_6 - \frac{1}{4l} \right) \tilde{d}^2(t)$$
$$+ \frac{1}{\sigma_2} \varepsilon_u^2 - \frac{1}{2}\sigma_1 \tilde{W}^T \tilde{W} + \frac{1}{2}\sigma_1 W^{*T} W^* + \frac{1}{\sigma_6} \bar{d}_v^2 \qquad (7.34)$$

Theorem 7.2 *Assume that the system (7.1)–(7.5) satisfies Assumption 7.1. Using the proposed boundary control law (7.31), RBF neural network updating law (7.20), and the observer (7.18), the closed loop system is globally stable and the following properties hold.*

Table 7.1 Parameters of a flexible hose system

Parameter	Description	Value
L	Length of the hose	16 m
ρ	Mass per unit length of the hose	5.2 kg/m
m	Mass of the drogue	39.5 kg
g	Acceleration of gravity	9.8 m/s^2
ρ_{air}	Air density	0.909 kg/m^3
D	Diameter of the hose	0.066 m
D_{drog}	Diameter of the drogue	0.61 m
C_f	Skin friction coefficient	0.005
C_d	Pressure drag coefficient	0.45
C_{drog}	Drag coefficient	0.43
b	Bound of output constraint	0.08

(1) The UB and UUB of the flexible hose system are proven;

(2) If the initial condition $w(L, 0)$ satisfies $|w(L, 0)| < b$, then $w(L, t)$ will remain in the constrained space, i.e., $|w(L, t)| < b$.

Proof See Appendix 2.

7.5 Simulation

For the purpose of illustrating the system performance, we utilize a finite difference method to compute numerical solution. And the simulations are implemented to demonstrate the validity of the presented control scheme. The disturbance $d(t)$ is given as $d(t) = 1.5 + sin(t)$. The initial displacement and velocity are given as $w(x, 0) = 2.4 \times 10^{-4}x^2$ and $w_t(x, 0) = 0$. The unknown deadzone is defined as $m_r = m_l = 1$, $b_r = 6$ and $b_l = -10$. The parameters of the flexible hose are listed in Table 7.1.

In order to analyze and verify the control effect, the dynamic responses of the flexible hose system are simulated in the following two cases:

Case 1: Without control: $u(t) = 0$,

Case 2: Applying the proposed control (7.31) and (7.19) with the same control gains $c_1 = 40$, $c_2 = 20$ and $\beta = 0.01$, and the parameters $\gamma_h = 3.5$ and $\sigma_1 = 0.02$ at speed $v(t) = 100$ m/s.

The dynamic responses for case 1 are shown in Fig. 7.3. It is clear that the deflection of the hose is large and the end-point deflection $w(L, t)$ transgresses its barrier.

The simulation results for case 2 are shown in Figs. 7.4, 7.5, 7.6, 7.7 and 7.8.

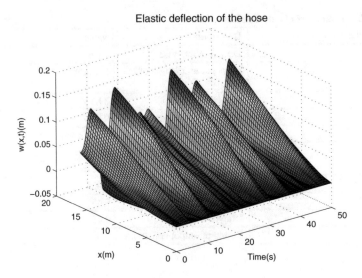

Fig. 7.3 Elastic deflection of the hose for case 1

The displacement $w(x, t)$ is shown in Fig. 7.4. Figure 7.5 depicts end-point displacement of the hose. From Figs. 7.4 and 7.5, we can see that both the proposed control schemes (7.19) and (7.31) can refrain the displacement greatly within 10 s. Therefore, the perfect control performance can be acquired with the presented control scheme even if there are input deadzone and external disturbances. Figure 7.6 shows the control input signal. Comparing the two sub-figures in Fig. 7.6a, it can be seen that when $b_l \leq u_0 \leq b_r$, where $b_l = -10$, $b_r = 6$ and $D(u_0) = 0$, the stability can also be obtained.

However, form Fig. 7.7b, we can see that the end-point displacement of the hose will transgress its barrier when use the basic control scheme (7.19) with the very small control gains c_1 and c_2. Using the control scheme (7.31) with the various control gains c_1 and c_2, $w(L, t)$ does not transgress its barrier, that is $w(L, t)$ always satisfies $|w(L, t)| < b$, as shown in Fig. 7.7a. Moreover, as the control gain increases, the end-point displacement of the hose $w(L, t)$ converges to zero at a faster rate and with less oscillation.

From Fig. 7.8 we can see that the observation error of the disturbance observer converges to zero. Using a disturbance observer rather than a robust control method has an advantage in the safe use of the actuator.

From above analysis of the simulations, we can come to a conclusion that the validity of the control strategy proposed in this paper can be guaranteed in handing the input deadzone, the external disturbances and the output constraint.

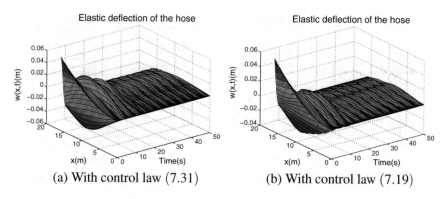

Fig. 7.4 Elastic deflection of the hose for case 2

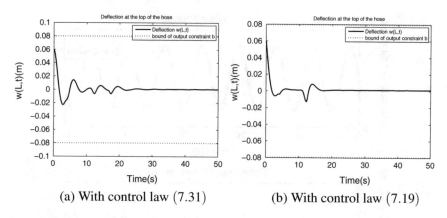

Fig. 7.5 End-point deflection of the hose for case 2

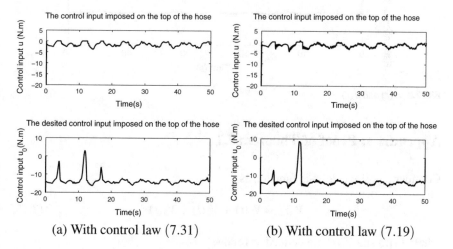

Fig. 7.6 The deadzone and desired control input for case 2

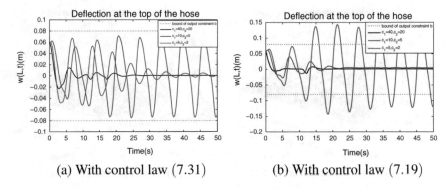

(a) With control law (7.31)   (b) With control law (7.19)

Fig. 7.7 End-point deflection of the hose $w(L, t)$ for various values of control gains

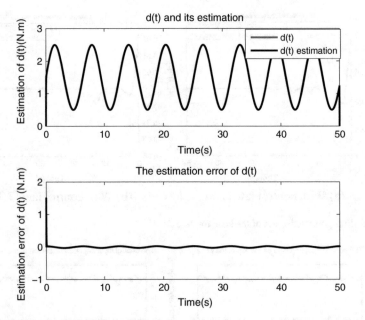

Fig. 7.8 $\hat{d}(t)$ and $\tilde{d}(t)$

Appendix 1: Proof of Theorem 7.1

Consider a positive definite function as

$$V(t) = V_1(t) + V_2(t) + V_b(t) \tag{7.35}$$

where $V_1(t)$ and $V_2(t)$ are defined as follows

$$V_1(t) = \frac{\beta}{2} \int_0^L \rho w_t^2(x,t) dx + \frac{\beta}{2} \int_0^L P(x,t)[w_x(x,t)]^2 dx$$

$$V_2(t) = \alpha\rho \int_0^L x w_t(x,t) w_x(x,t) dx$$

where $\alpha > 0$ and $\beta > 0$.

We can see $V_2(t)$ satisfies

$$|V_2(t)| \le \alpha\rho L \int_0^L w_t^2(x,t) dx + \alpha\rho L \int_0^L [w_x(x,t)]^2 dx \le \beta_1 V_1(t)$$

where $\beta_1 = \frac{2\alpha\rho L}{\beta \min(\rho, P_{\min})}$. We then obtain

$$-\beta_1 V_1(t) \le V_2(t) \le \beta_1 V_1(t)$$

Assuming that α satisfies $0 < \alpha < \frac{\beta \min(\rho, P_{\min})}{2\rho L}$, we can obtain $0 < \beta_1 < 1$, and

$$\beta_2 (V_1(t) + V_b(t)) \le V(t) \le \beta_3 (V_1(t) + V_b(t)) \tag{7.36}$$

where $\beta_2 = \min(1 - \beta_1, 1) = 1 - \beta_1$ and $\beta_3 = \max(1 + \beta_1, 1) = 1 + \beta_1$.

Differentiating Eq. (7.35) with respect to time, then combining Eq. (7.1) and boundary equations, using integration by parts, and according to Lemmas 2.4 and 2.5, we obtain

$$
\begin{aligned}
\dot{V} &= \dot{V}_1 + \dot{V}_2 + \dot{V}_b \\
&\le -\frac{1}{2} \int_0^L \left[\alpha P(x,t) - \alpha x P'(x,t) \right] [w_x(x,t)]^2 dx \\
&\quad - \frac{1}{2} \int_0^L \left[-2\alpha L \sigma_5 - \beta \dot{P}(x,t) \right] [w_x(x,t)]^2 dx \\
&\quad - \left(\frac{\alpha\rho}{2} - \frac{\beta}{\sigma_4} \right) \int_0^L w_t^2(x,t) dx - \frac{1}{2}\sigma_1 \tilde{W}^T \tilde{W} \\
&\quad - (\beta P(L,t) - \alpha\rho L) \frac{w_t^2(L,t)}{2} - \left(K - \sigma_6 - \frac{1}{4l} \right) \tilde{d}^2(t) \\
&\quad - \left(\frac{\beta P(L,t)}{2} - \frac{\alpha}{2} L P(L,t) - \sigma_3 \right) [w_x(L,t)]^2 \\
&\quad - \left(c_1 - \frac{\beta P(L,t) c_1^2}{2} - \frac{1}{\sigma_3} \right) z_1^2 + \frac{1}{\sigma_2}\varepsilon_u^2 \\
&\quad - \left(c_2 - \sigma_2 - \frac{\beta P(L,t)}{2} \right) z_2^2 + \frac{1}{2}\sigma_1 W^{*T} W^*
\end{aligned}
$$

$$+ \left(\frac{\alpha L}{\sigma_5} + \sigma_4 \beta \right) \int_0^L Q^2(x, t) dx + \frac{1}{\sigma_6} \bar{d}_v^2 \tag{7.37}$$

We design parameters α and β to satisfy the following inequality:

$$\alpha P_{\min} - \beta \dot{P}_{\max} - \alpha L P'_{\max} - 2\alpha L \sigma_5 \geq \delta$$

$\forall (x, t) \in [0, L] \times [0, \infty)$, for a positive constant δ, and the following conditions:
$\frac{\alpha \rho}{2} - \frac{\beta}{\sigma_4} > 0$, $c_1 - \frac{\beta P_{\max} c_1^2}{2} - \frac{1}{\sigma_3} > 0$, $c_2 - \sigma_2 - \frac{\beta P_{\max}}{2} \geq 0$, $\frac{\beta P_{\min}}{2} - \frac{\alpha}{2} L P_{\max} - \sigma_3 \geq 0$,
$K - \sigma_6 - \frac{1}{4l} > 0$, $\beta P_{\min} - \alpha \rho L \geq 0$.
Equation (7.37) can be rewritten as

$$\dot{V}(t) \leq -\gamma_1 \frac{\beta}{2} \int_0^L P(x, t)[w_x(x, t)]^2 dx - \gamma_2 \frac{\beta \rho}{2} \int_0^L w_t^2(x, t) dx$$
$$- \gamma_3 \frac{1}{2} z_1^2 - \gamma_4 \frac{m}{2} z_2^2 - \gamma_5 \frac{1}{2\gamma_h} \tilde{W}^T \tilde{W} - \gamma_6 \frac{1}{2} \tilde{d}^2(t) + \varepsilon \tag{7.38}$$

where $\gamma_1 = \frac{\delta}{\beta P_{\max}}$, $\gamma_2 = \left(\frac{\alpha}{\beta} - \frac{2}{\sigma_4 \rho} \right)$, $\gamma_3 = 2 \left(c_1 - \frac{\beta P_{\max} c_1^2}{2} - \frac{1}{\sigma_3} \right)$, $\gamma_4 = \frac{2}{m} \left(c_2 - \sigma_2 - \frac{\beta P_{\max}}{2} \right)$, $\gamma_5 = \gamma_h \sigma_1$, $\gamma_5 = 2 \left(K - \sigma_6 - \frac{1}{4l} \right)$,
$\varepsilon = \left(\frac{\alpha L}{\sigma_5} + \sigma_4 \beta \right) L Q_{\max}^2 + \frac{1}{\sigma_2} \varepsilon_u^2 + \frac{1}{2} \sigma_1 W^{*T} W^* + \frac{1}{\sigma_6} \bar{d}_v^2$.
We further obtain

$$\dot{V}(t) \leq -\lambda_1 [V_1(t) + V_b(t)] + \varepsilon \tag{7.39}$$

where $\lambda_1 = \min (\gamma_1, \gamma_2, \gamma_3, \gamma_4, \gamma_5, \gamma_6)$.
Combining (7.36) and (7.39) yields

$$\dot{V}(t) \leq -\lambda V(t) + \varepsilon \tag{7.40}$$

where $\lambda = \lambda_1 / \beta_3 > 0$.
Then multiplying Eq. (7.40) by $e^{\lambda t}$, and integrating of the inequality, we have

$$V(t) \leq V(0) e^{-\lambda t} + \frac{\varepsilon}{\lambda} \left(1 - e^{-\lambda t} \right)$$

According to Lemma 2.5, we have

$$\frac{\beta P_{\min}}{2L} w^2(x, t) \leq \frac{\beta}{2} \int_0^L P(x, t)[w_x(x, t)]^2 dx \leq V_1(t)$$
$$\leq V_1(t) + V_b(t) \leq \frac{V(t)}{\beta_2} \tag{7.41}$$

Then we deduce that $w(x, t)$ satisfies

$$|w(x, t)| \leq \sqrt{\frac{2L}{\beta_2 \beta P_{min}} \left[V(0) e^{-\lambda t} + \frac{\varepsilon}{\lambda} \left(1 - e^{-\lambda t} \right) \right]} \leq C_1$$

We finally obtain, $\lim\limits_{t \to \infty} |w(x, t)| \leq \sqrt{\frac{2L\varepsilon}{\beta P_{min} \beta_2 \lambda}} = C_2, \forall (x, t) \in [0, L] \times [0, \infty)$, so $w(x, t)$ is uniformly ultimate bounded.

Appendix 2: Proof of Theorem 7.2

Choosing the positive definite function as

$$\bar{V}(t) = V_1(t) + V_2(t) + \bar{V}_b(t) \tag{7.42}$$

its time derivative is

$$\dot{\bar{V}}(t) = \dot{V}_1(t) + \dot{V}_2(t) + \dot{\bar{V}}_b(t)$$
$$\leq -\frac{1}{2} \int_0^L \left[\alpha P(x, t) - \alpha x P'(x, t)) \right] [w_x(x, t)]^2 dx$$
$$- \frac{1}{2} \int_0^L \left[-2\alpha L\sigma_5 - \beta \dot{P}(x, t) \right] [w_x(x, t)]^2 dx$$
$$- \left(c_1 - \frac{\beta P(L, t) c_1^2}{2} - \frac{1}{\sigma_3} \right) \frac{z_1^2}{b^2 - z_1^2}$$
$$- \left(\frac{\alpha \rho}{2} - \frac{\beta}{\sigma_4} \right) \int_0^L w_t^2(x, t) dx$$
$$- \frac{\beta P(L, t)}{2} (b^2 - z_1^2) [w_x(L, t)]^2$$
$$+ \left(\frac{\alpha}{2} L P(L, t) + \sigma_3 \right) (b^2 - z_1^2) [w_x(L, t)]^2$$
$$- \left(\beta P(L, t) - \alpha \rho L (b^2 - z_1^2) \right) \frac{w_t^2(L, t)}{2(b^2 - z_1^2)}$$
$$- (c_2 - \sigma_2) z_2^2 - \left(K - \sigma_6 - \frac{1}{4l} \right) \tilde{d}^2(t)$$
$$+ \frac{1}{\sigma_2} \varepsilon_u^2 - \frac{1}{2} \sigma_1 \tilde{W}^T \tilde{W} + \frac{1}{2} \sigma_1 W^{*T} W^*$$
$$+ \left(\frac{\alpha L}{\sigma_5} + \sigma_4 \beta \right) \int_0^L Q^2(x, t) dx + \frac{1}{\sigma_6} \tilde{d}_v^2 \tag{7.43}$$

The parameters α and β are designed to satisfy

$$\alpha P_{min} - \beta \dot{P}_{max} - \alpha L P'_{max} - 2\alpha L \sigma_5 \geq \delta,$$

for a positive constant δ, and the following conditions:

$\frac{\alpha\rho}{2} - \frac{\beta}{\sigma_4} > 0$, $c_1 - \frac{\beta P_{max} c_1^2}{2} - \frac{1}{\sigma_3} > 0$, $c_2 - \sigma_2 \geq 0$, $\frac{\beta P_{min}}{2} - \frac{\alpha}{2} L P_{max} - \sigma_3 \geq 0$,
$K - \sigma_6 - \frac{1}{4l} > 0$, $\beta P_{min} - \alpha\rho L b^2 \geq 0$.

Equation (7.43) then can be rewritten as

$$\dot{V}(t) \leq -\gamma_1 \frac{\beta}{2} \int_0^{l} P(x, t)[w_x(x, t)]^2 dx - \gamma_2 \frac{\beta\rho}{2} \int_0^L w_t^2(x, t) dx$$
$$- \gamma_3 \frac{1}{2} \ln \frac{b^2}{b^2 - z_1^2} - \gamma_4 \frac{m}{2} z_2^2 - \gamma_5 \frac{1}{2\gamma_h} \tilde{W}^T \tilde{W} - \gamma_6 \frac{1}{2} \tilde{d}^2(t) + \varepsilon \qquad (7.44)$$

where $\gamma_4 = \frac{2}{m}(c_2 - \sigma_2)$, and γ_1, γ_2, γ_3, γ_5, γ_6 and ε are the same as in the proof of Theorem 7.1.

The remaining of the proof is omitted since it is similar to the proof of Theorem 7.2.

Equations (7.36) and (7.33) indicates $\bar{V}_b(t)$ is nonnegative and bounded $\forall t \in [0, \infty)$. From the definition of $\bar{V}_b(t)$, we know that $\bar{V}_b(t) \to \infty$, as $|w(L, t)| \to b$. Given that $|w(L, 0)| < b$, and according to Lemma 1 presented in [6], we infer that $w(L, t)$ satisfies $|w(L, t)| < b$.

References

1. He W, Ge SS (2015) Vibration control of a flexible beam with output constraint. IEEE Trans Ind Electron 62(8):5023–5030
2. He W, Ge SS (2016) Cooperative control of a nonuniform gantry crane with constrained tension. Automatica 66:146–154
3. He W, Ouyang Y, Hong J (2017) Vibration control of a flexible robotic manipulator in the presence of input deadzone. IEEE Trans Ind Inform 13(1):48–59
4. He W, Sun C, Ge SS (2015) Top tension control of a flexible marine riser by using integral-barrier Lyapunov function. IEEE/ASME Trans Mechatron 20(2):497–505
5. He W, Zhang S, Ge SS (2014) Adaptive control of a flexible crane system with the boundary output constraint. IEEE Trans Ind Electron 61(8):4126–4133
6. Ren B, Ge SS, Tee KP, Lee TH (2010) Adaptive neural control for output feedback nonlinear systems using a barrier Lyapunov function. IEEE Trans Neural Netw 21(8):1339–1345
7. Shyu K-K, Liu W-J, Hsu K-C (2005) Design of large-scale time-delayed systems with dead-zone input via variable structure control. Automatica 41(7):1239–1246
8. Tee KP, Ren B, Ge SS (2011) Control of nonlinear systems with time-varying output constraints. Automatica 47(11):2511–2516
9. Wang X-S, Su C-Y, Henry H (2004) Robust adaptive control of a class of nonlinear systems with unknown dead-zone. Automatica 40(3):407–413
10. Yang Q, Chen M (2015) Adaptive neural prescribed performance tracking control for near space vehicles with input nonlinearity. Neurocomputing 80(3):1509–1520

Chapter 8
Modeling and Vibration Control of a Flexible Aerial Refueling Hose with Variable Lengths and Input Constraint

8.1 Introduction

In Chaps. 5–7, we study the modeling and control design for a flexible aerial refueling hose system during coupling or before coupling but with fixed length of the hose. In practice, the elongation of the hose will seriously affect the dynamics and control performance, which will lead to premature fatigue failure. In this chapter, we will present boundary control design for a flexible aerial refueling hose with varying length, horizontal and vertical speeds. Modeling and control of moving systems have typically been studied [2, 6, 7]. In [5], boundary control schemes are proposed to reduce the vibrations of a stretched moving string, where an iterative learning algorithm is used to deal with the effects of the external boundary disturbance. In [3], an active control scheme for an axially moving string system that suppresses both longitudinal and transverse vibrations and regulates the transport velocity of the string to track a desired moving velocity profile is investigated. In [4], a novel control algorithm is discussed for an axially moving membrane system to regulate the transverse vibrations and track a desired axial transport velocity. In [1], boundary control laws are developed to stabilize the transverse vibration for a nonlinear vertically moving string system which is considered with varying length, varying speed, and the constrained boundary output. The above works mainly deal with the problems for moving systems with a horizontal or a vertical speed, however, the flexible aerial refueling hose system have both horizontal and vertical speeds.

In this chapter, we study boundary control design for a flexible aerial refueling hose with varying length, horizontal and vertical speeds, and input constraint. The flexible hose is described by PDEs. With the backstepping technology, a novel boundary controller is proposed for the flexible hose based on PDEs. With the proposed control, the closed-loop stability is proved based on the Lyapunv's direct method and the deflection eventually converges to an arbitrarily small neighborhood around the origin. The main contributions of this chapter are that: (1) a boundary control scheme with a smooth hyperbolic function is proposed to suppress the vibration of the flexible hose subject to varying length, varying speed, and input constraint, (2) an

© Tsinghua University Press 2020
Z. Liu and J. Liu, *PDE Modeling and Boundary Control for Flexible Mechanical System*, Springer Tracts in Mechanical Engineering,
https://doi.org/10.1007/978-981-15-2596-4_8

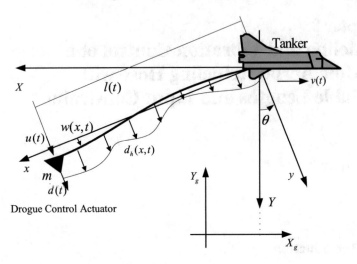

Fig. 8.1 Diagram of a flexible aerial refueling hose

auxiliary system is designed to compensate for the nonlinear term arising from the input saturation, and (3) the close-loop stability analysis avoids any simplification or discretization of the PDEs based on the Lyapunov direct method.

The rest of the chapter is organized as follows. The PDE dynamic model of a flexible aerial refueling hose is derived in Sect. 8.2. In Sect. 8.3, a boundary control scheme is designed. In Sect. 8.4, the closed-loop stability is proved based on the Lyapunov direct method. Numerical simulations are demonstrated in Sect. 8.5 to show the effectiveness of the proposed controller.

8.2 Problem Formulation

In this chapter, the dynamical model of the tanker is not considered, and the tanker provides the same speed for the undeformed hose, as shown in Fig. 8.1.

The kinetic energy of the hose system including the drogue $E_k(t)$ can be represented as

$$
\begin{aligned}
E_k(t) = \frac{\rho}{2} \int_0^{l(t)} & \left\{ \left(\frac{Dz_x(x,t)}{Dt} \right)^2 + \left(\frac{Dz_y(x,t)}{Dt} \right)^2 \right\} dx \\
+ \frac{1}{2} m & \left\{ \left(\frac{Dz_x(l(t),t)}{Dt} \right)^2 + \left(\frac{Dz_y(l(t),t)}{Dt} \right)^2 \right\}
\end{aligned}
\tag{8.1}
$$

where x represents the independent spatial variable $0 \le x \le l(t) \le L$ and t represents the independent time variable, ρ is mass per unit length of the hose, $l(t)$ and L are time-varying length and total length of the hose, m is mass of the drogue. The material derivative of a moving material is defined by $D(*)/Dt = \partial(*)/\partial t + v_m\partial(*)/\partial x$, v_m is the speed of the moving material.

The gravitational potential energy of the hose system $E_{pg}(t)$ can be expressed as

$$E_{pg}(t) = \rho g \int_0^{l(t)} z_y(x, t)dx + mgz_y(l(t), t) \tag{8.2}$$

and the potential energy of the hose system $E_{pf}(t)$ due to the axial force can be obtained from

$$E_{pf}(t) = \frac{1}{2}\int_0^{l(t)} P(x, t)[w_x(x, t)]^2 dx \tag{8.3}$$

where $P(x, t)$ is the tension of the hose that can be expressed as [7]

$$P(x, t) = [m + \rho(l(t) - x)](g\sin\theta - l_{tt}(t) - r_{tt}(t)\cos\theta)$$
$$+ f_{drog}(t)\cos\theta + f_t(x, t) \tag{8.4}$$

where g is the acceleration of gravity, $f_t(x, t)$ is the skin friction drag in the tangential direction, and $f_{drog}(t)$ is the drag of the drogue. So the potential energy of the hose system is

$$E_p(t) = E_{pg}(t) + E_{pf}(t) \tag{8.5}$$

The virtual work done on the system is given by

$$\delta W(t) = \int_0^{l(t)} d_h(x, t)\delta w(x, t)\,dx - \int_0^{l(t)} f_n(x, t)\delta w(x, t)\,dx$$
$$+ \left(u(t) + d(t) - f_{drog}(t)\sin\theta\right)\delta w(l(t), t) \tag{8.6}$$

where $f_n(x, t)$ is the pressure drag in the normal direction, $d(t)$ is boundary disturbance on the actuator, and $d_h(x, t)$ is distributed disturbance on the hose.

Then, the Hamilton's principle is applied as

$$\int_{t_1}^{t_2} (\delta E_k(t) - \delta E_p(t) + \delta W(t))dt = 0 \tag{8.7}$$

We further obtain the following PDEs of the hose system as

$$\rho\left[w_{tt}(x, t) + 2l_t(t)w_{tx}(x, t) + l_{tt}(t)w_x(x, t) + l_t^2(t)w_{xx}(x, t)\right]$$
$$= P_x(x, t)w_x(x, t) + P(x, t)w_{xx}(x, t) + Q(x, t) \tag{8.8}$$

where

$$Q(x, t) = d_h(x, t) - f_n(x, t) + \rho(g\cos\theta - r_{tt}(t)\sin\theta) \tag{8.9}$$

and boundary conditions of the hose system as

$$mw_{tt}(l(t), t) + mr_{tt}(t) \sin \theta - mg \cos \theta + P(l(t), t)w_x (l(t), t)$$
$$+ m \left[2l_t(t)w_{tx}(x, t) + l_{tt}(t)w_x(x, t) + l_t^2(t)w_{xx}(x, t) \right]$$
$$= -f_{drog}(t) \sin \theta + u(t) + d(t) \tag{8.10}$$

$$w(0, t) = w_t(0, t) = 0 \tag{8.11}$$

Remark 8.1 It is noted that, related to suppressing the transversal vibration, only the vibration of $w(x, t)$ is considered. The position of the aircraft $r(t)$ and the hose length $l(t)$ are prespecified functions, therefore, it is not necessary to consider their variations, that is $\delta r(t) = \delta r_t(t) = 0$ and $\delta l(t) = \delta l_t(t) = 0$.

Assumption 8.1 The boundary disturbance $d(t)$ and the distributed disturbance $d_h(x, t)$ are bounded so that there exist two positive constants $\bar{d}_D$ and $\bar{d}_h$ satisfying $|d(t)| \leq \bar{d}_D$ and $|d_h(x, t)| \leq \bar{d}_h$.

Assumption 8.2 ([1]) For the positive definite function $P(x, t)$, we assume that $P(x, t)$, $P_x(x, t)$ and $P_t(x, t)$ are bounded by known, constant lower, and upper bounds as follows,

$$0 \leq P_{\min} \leq P(x, t) \leq P_{\max}$$

$$0 \leq P_{x\min} \leq P_x(x, t) \leq P_{x\max}$$

$$0 \leq P_{t\min} \leq P_t(x, t) \leq P_{t\max}$$

$\forall (x, t) \in [0, L] \times [0, \infty)$.

Assumption 8.3 We assume that the function $Q(x, t)$ is bounded so that there exist a positive constant $Q_{\max}$ satisfying $|Q(x, t)| \leq Q_{\max}$, $\forall (x, t) \in [0, L] \times [0, \infty)$.

Remark 8.2 From the definition of $f_n(x, t)$, $v_t(t)$, and Assumption 8.1, we can obtain the value of $Q(x, t)$ according to (8.9). If velocity $v(t)$ and acceleration $v_t(t)$ of the tanker are known, then it is possible to calculate the maximum value of $Q(x, t)$.

8.3 Control Design

The control objective is to propose a controller $u(t)$ to suppress the elastic vibration $w(x, t)$ of the flexible aerial refueling hose in the presence of the high speed of the aerial refueling hose system and input saturation. In this section, we use the backstepping method to design a boundary control law $u(t)$ on the top boundary of the hose and use Lyapunov's direct method to analyze the closed-loop stability of the system.

To achieve the objective, the input saturation model is described as follows

$$u(t) = u_g(u_0) = u_M \tanh\left(\frac{u_0}{u_M}\right) \tag{8.12}$$

where u_M is a known bound of $u(t)$ and u_0 is the designed control command. Then the hose system with input saturation are designed as follows:

$$\rho\left[w_{tt}(x,t) + 2l_t(t)w_{xt}(x,t) + l_{tt}(t)w_x(x,t) + l_t^2(t)w_{xx}(x,t)\right]$$
$$= P_x(x,t)w_x(x,t) + P(x,t)w_{xx}(x,t) + Q(x,t) \tag{8.13}$$

$$Q(x,t) = d_h(x,t) - f_n(x,t) + \rho\left(g\cos\theta - r_{tt}(t)\sin\theta\right) \tag{8.14}$$

$$mw_{tt}(l(t),t) + mr_{tt}(t)\sin\theta - mg\cos\theta + P(l(t),t)w_x(l(t),t)$$
$$+ m\left[2l_t(t)w_{xt}(x,t) + l_{tt}(t)w_x(l(t),t) + l_t^2(t)w_{xx}(l(t),t)\right]$$
$$= -f_{drog}(t)\sin\theta + u_g(u_0) + d(t) \tag{8.15}$$

$$\dot{u}_0 = -cu_0 + \omega \tag{8.16}$$

where $c > 0$ and ω is an auxiliary signal to be designed in the following backstepping approach.

As the usual backstepping approach, the following transform of coordinate is made:

$$z_1 = x_1 = w(l(t),t)$$
$$z_2 = x_2 - \tau_1 = \frac{Dx_1}{Dt} - \tau_1$$
$$= w_t(l(t),t) + l_t(t)w_x(l(t),t) - \tau_1$$
$$z_3 = u_g(u_0) - \tau_2 \tag{8.17}$$

where τ_1 and τ_2 are the virtual control determined at first and second step that

$$\tau_1 = -c_1z_1 - w_x(l(t),t) \tag{8.18}$$

$$\tau_2 = -(c_2 + c_l)z_2 - z_1 + \beta c_1 P(l(t),t)z_1 - mg\cos\theta$$
$$+ f_{drog}(t)\sin\theta + P(l(t),t)w_x(l(t),t) + mr_{tt}(t)\sin\theta + m\dot{\tau}_1 \tag{8.19}$$

where c_1, c_2 and c_l are positive constants, and $\dot{\tau}_1 = D\tau_1/Dt$.

Step 1. We consider a Lyapunov function candidate as

$$V_{b1} = \frac{1}{2}z_1^2 \tag{8.20}$$

The material derivative of (8.20) is

$$\dot{V}_{b1} = z_1 z_{t1} = z_1 x_2 = z_1 (z_2 + \tau) \tag{8.21}$$

Substituting (8.18) in (8.21), we have

$$\dot{V}_{b1} = -c_1 z_1^2 - z_1 w_x(l(t), t) + z_1 z_2 \tag{8.22}$$

Step 2. Similarly, consider a Lyapunov function candidate as

$$V_{b2} = V_{b1} + \frac{1}{2}m z_2^2 \tag{8.23}$$

Noting that $z_3 = u_g(u_0) - \tau_2$, using (8.19) and Lemma 2.4, we obtain the material derivative of V_{b2}

$$\dot{V}_{b2} \le -c_1 z_1^2 - c_2 z_2^2 - z_1 w_x (l(t), t) + z_2 z_3 + \frac{1}{4c_l}d^2(t)$$
$$+ \beta c_1 P(l(t), t) z_1 x_2 + \beta c_1 P(l(t), t) c_1 z_1^2$$
$$+ \beta c_1 P(l(t), t) z_1 w_x(l(t), t) \tag{8.24}$$

Step 3. Considering (8.16), (8.17) and (8.19), we obtain

$$m\dot{z}_3 = m \frac{\partial u_g}{\partial u_0}(-cu_0 + \omega) - m\dot{\tau}_2$$
$$= m\xi(-cu_0 + \omega) - \frac{\partial \tau_2}{\partial x_2}d(t) - \vartheta \tag{8.25}$$

$$\vartheta = \frac{\partial \tau_2}{\partial x_2}\left(u_g(u_0) - P(l(t), t)w_x(l(t), t) - mr_{tt}(t)\sin\theta\right)$$
$$+ \frac{\partial \tau_2}{\partial x_2}\left(mg\cos\theta - f_{drog}(t)\sin\theta\right) + m\frac{\partial \tau_2}{\partial x_1}x_2$$
$$+ m\frac{\partial \tau_2}{\partial P(l(t), t)}P_t(l(t), t) + m\frac{\partial \tau_2}{\partial f_{drog}(t)}\dot{f}_{drog}(t)$$
$$+ m\frac{\partial \tau_2}{\partial w_x(l(t), t)}\frac{Dw_x(l(t), t)}{Dt}$$
$$+ m\frac{\partial \tau_2}{\partial w_{xt}(l(t), t)}\frac{Dw_{xt}(l(t), t)}{Dt}$$

$$+ m \frac{\partial \tau_2}{\partial w_{xx}(l(t), t)} \frac{Dw_{xx}(l(t), t)}{Dt} \tag{8.26}$$

$$\xi = \frac{\partial u_g(u_0)}{\partial u_0} = \frac{4}{\left(e^{u_0/u_M} + e^{-u_0/u_M}\right)^2} > 0 \tag{8.27}$$

Then we use a Nussbaum function $N(\chi)$ to design the control law ω in (8.16), and ω is designed as

$$\omega = N(\chi)\bar{\omega} \tag{8.28}$$

$$\bar{\omega} = \xi c u_0 + \frac{1}{m}\vartheta - \frac{1}{m}c_3 z_3 - \frac{1}{m}z_2 - \frac{1}{m}c_l \left(\frac{\partial \tau_2}{\partial x_2}\right)^2 z_3 \tag{8.29}$$

where $N(\chi)$ is a Nussbaum function defined as

$$N(\chi) = \chi^2 \cos(\chi) \tag{8.30}$$

$$\dot{\chi} = \gamma_\chi m z_3 \bar{\omega} \tag{8.31}$$

where γ_χ is a positive real design parameter.

We choose a Lyapunov function candidate

$$V_b = V_{b2} + \frac{1}{2}m z_3^2 \tag{8.32}$$

The material derivative of V_b is

$$\begin{aligned} \dot{V}_b \leq &-c_1 z_1^2 - c_2 z_2^2 - z_1 w_x(l(t), t) + \frac{1}{4c_d}d^2(t) \\ &+ z_2 z_3 + z_3 (m z_{t3} + m\bar{\omega}) - m z_3 \bar{\omega} + \beta c_1 P(l(t), t) z_1 x_2 \\ &+ \beta c_1 P(l(t), t) c_1 z_1^2 + \beta c_1 P(l(t), t) z_1 w_x(l(t), t) \end{aligned} \tag{8.33}$$

Substituting (8.25) and (8.29) in (8.33), using (8.28), (8.31) and Lemma 2.4, we obtain

$$\begin{aligned} \dot{V}_b \leq &-c_1 z_1^2 - c_2 z_2^2 - c_3 z_3^2 + \beta c_1 P(l(t), t) c_1 z_1^2 \\ &+ \beta c_1 P(l(t), t) z_1 x_2 + \beta c_1 P(l(t), t) z_1 w_x(l(t), t) \\ &+ \frac{1}{\sigma_3}z_1^2 + \sigma_3[w_x(l(t), t)]^2 + \frac{1}{2c_l}\bar{d}_D^2 \\ &+ \frac{1}{\gamma_\chi}(\xi N(\chi) - 1)\dot{\chi} \end{aligned} \tag{8.34}$$

where σ_3 is a positive constant.

8.4 Stability Analysis

The stability of the closed-loop system is proven by Lyapunov theory.

Theorem 8.1 *Suppose the system (8.8)–(8.11) satisfies Assumptions 8.1–8.3. With the proposed boundary control law (8.12), (8.16), (8.18), (8.19), (8.28) and (8.29), the closed-loop system is globally stable and the following properties hold.*
(1) The control input is bounded, and its boundary is described as:

$$|u(t)| = \left| u_g\left(u_0\right) \right| = u_M \left| \tanh\left(\frac{u_0}{u_M}\right) \right| \leq u_M \tag{8.35}$$

(2) The closed-loop system is uniformly bounded and $w(x,t)$ will remain in the compact set Ω defined by

$$\Omega = \{ w(x,t) \in R : |w(x,t)| \leq C_1, \forall x \in [0, l(t)] \} \tag{8.36}$$

where $C_1 = \sqrt{\frac{2l(t)}{\beta P_{\min}\beta_2}\left(V(0) + \frac{\varepsilon_0}{\lambda}\right)}$ the parameters β, β_2, λ and ε_0 are defined in the following process of proof.

Proof Consider the Lyapunov candidate function as

$$V(t) = V_1(t) + V_2(t) + V_b(t) \tag{8.37}$$

where $V_1(t)$ and $V_2(t)$ are defined as follows

$$V_1(t) = \frac{\beta}{2} \int_0^{l(t)} \rho\left(\frac{Dw(x,t)}{Dt}\right)^2 dx$$
$$+ \frac{\beta}{2} \int_0^{l(t)} P(x,t)[w_x(x,t)]^2 dx \tag{8.38}$$

$$V_2(t) = \alpha \int_0^{l(t)} \rho x \left(w_t(x,t) + l_t(t)w_x(x,t)\right) w_x(x,t) dx \tag{8.39}$$

where α and β are positive weighting constants.

To motivate the followings, we first focus our attention on the term $V_2(t)$. It satisfies the following inequality

$$|V_2(t)| \leq \alpha\rho l(t) \int_0^{l(t)} \left(\frac{Dw(x,t)}{Dt}\right)^2 dx$$
$$+ \alpha\rho l(t) \int_0^{l(t)} [w_x(x,t)]^2 dx \leq \beta_1 V_1(t) \tag{8.40}$$

where $\beta_1 = \frac{2\alpha\rho L}{\beta \min(\rho, P_{\min})}$. Assuming that α is a small positive weighting constant satisfying $0 < \alpha < \frac{\beta \min(\rho, P_{\min})}{2\rho L}$, we can obtain $0 < \beta_1 < 1$, and

$$\beta_2 (V_1(t) + V_b(t)) \le V(t) \le \beta_3 (V_1(t) + V_b(t)) \tag{8.41}$$

where $\beta_2 = \min(1 - \beta_1, 1) = 1 - \beta_1$ and $\beta_3 = \max(1 + \beta_1, 1) = 1 + \beta_1$.

The material derivative of (8.37) is

$$\dot{V}(t) = \dot{V}_1(t) + \dot{V}_2(t) + \dot{V}_3(t) \tag{8.42}$$

Applying (8.8) and boundary equations, according to Lemma 2.4, we can obtain

$$
\begin{aligned}
\dot{V}_1(t) \le{}& \frac{\beta}{2\sigma_1} \int_0^{l(t)} [w_t(x, t) + l_t(t) w_x(x, t)]^2 dx \\
&+ \frac{1}{2}\beta P(l(t), t) z_2^2 + \frac{1}{2}\beta \int_0^{l(t)} P_t(x, t)[w_x(x, t)]^2 dx \\
&- \beta c_1 P(l(t), t) [w_t(l(t), t) + l_t(t) w_x(l(t), t)] w(l(t), t) \\
&- \frac{1}{2}\beta P(l(t), t) [w_t(l(t), t) + l_t(t) w_x(l(t), t)]^2 \\
&- \frac{1}{2}\beta (P(l(t), t) + l_t(t) P(l(t), t)) [w_x(l(t), t)]^2 \\
&- \beta c_1 P(l(t), t) w_x(l(t), t) w(l(t), t) \\
&- \frac{1}{2}\beta \int_0^L l_t(t) P_x(x, t)[w_x(x, t)]^2 dx \\
&- \frac{1}{2}\beta P(l(t), t) c_1^2 w^2(l(t), t) \\
&- \frac{1}{2}\beta l_t(t) P(0, t)[w_x(0, t)]^2 \\
&+ \frac{\sigma_1 \beta}{2} \int_0^{l(t)} Q^2(x, t) dx
\end{aligned}
\tag{8.43}
$$

where σ_1 is a positive constant.

To go on, we concentrate our attention on $\dot{V}_2(t)$. Substituting (8.8), using integration by parts, the boundary conditions and Lemma 2.4, we obtain

$$
\begin{aligned}
\dot{V}_2(t) \le{}& \alpha \int_0^{l(t)} \left(x P_x(x, t) - \frac{1}{2} P(x, t) \right) [w_x(x, t)]^2 dx \\
&- \frac{\alpha}{2} \int_0^{l(t)} (x P_x(x, t) - \sigma_2 l(t)) [w_x(x, t)]^2 dx \\
&- \frac{1}{2}\alpha\rho \int_0^{l(t)} [w_t(x, t) + l_t(t) w_x(x, t)]^2 dx
\end{aligned}
$$

$$+ \frac{1}{2} \alpha \rho [w_x (l(t), t)]^2 + \frac{\alpha l(t)}{2\sigma_2} \int_0^{l(t)} Q^2(x, t) dx \tag{8.44}$$

where σ_2 is a positive constant.

Substituting (8.34), (8.43), and (8.44) into (8.42), we obtain

$$\dot{V}(t) = \dot{V}_1(t) + \dot{V}_2(t) + \dot{V}_3(t)$$

$$\leq -\frac{1}{2} \int_0^{l(t)} [\alpha P(x, t) - \alpha x P_x(x, t) - \beta P_t(x, t)] [w_x (x, t)]^2 dx$$

$$-\frac{1}{2} \int_0^{l(t)} [\beta l_t(t) P_x(x, t) - \alpha \sigma_2 l(t)] [w_x (x, t)]^2 dx$$

$$-\frac{1}{2} \left(\alpha \rho - \frac{\beta}{\sigma_1} \right) \int_0^{l(t)} [w_t(x, t) + l_t(t) w_x (x, t)]^2 dx$$

$$-\frac{1}{2} (\beta P (l(t), t) + \beta l_t(t) P(l(t), t) - \alpha \rho - 2\sigma_3) [w_x (l(t), t)]^2$$

$$-\frac{1}{2} \beta P (l(t), t) [w_t (l(t), t) + l_t(t) w_x (l(t), t)]^2$$

$$-\left(c_1 - \frac{1}{2} \beta c_1^2 P (l(t), t) - \frac{1}{\sigma_3} \right) z_1^2 - \left(c_2 - \frac{1}{2} \beta P (l(t), t) \right) z_2^2$$

$$- c_3 z_3^2 - \frac{1}{2} \beta l_t(t) P(0, t) [w_x (0, t)]^2 + \frac{1}{\gamma_\chi} (\xi N (\chi) - 1) \dot{\chi}$$

$$+ \frac{1}{2c_l} \bar{d}_D^2 + \frac{1}{2} \left(\sigma_1 \beta + \frac{\alpha l(t)}{\sigma_2} \right) \int_0^{l(t)} Q^2(x, t) dx \tag{8.45}$$

We design parameters α and β to satisfy the following inequality:

$$\alpha P_{\min} + \beta l_t(t) P_{x\min} - \alpha L P_{x\max} - \beta P_{t\max} - \alpha \sigma_2 l(t) \geq \delta \tag{8.46}$$

$\forall (x, t) \in [0, L] \times [0, \infty)$, for a positive constant δ, and the following conditions:

$$\alpha \rho - \frac{\beta}{\sigma_1} \geq 0 \tag{8.47}$$

$$c_1 - \frac{1}{2} \beta c_1^2 P_{\max} - \frac{1}{\sigma_3} \geq 0 \tag{8.48}$$

$$c_2 - \frac{1}{2} \beta P_{\max} \geq 0 \tag{8.49}$$

$$(1 + l_t(t)) \beta P_{\min} - \alpha \rho - 2\sigma_3 \geq 0 \tag{8.50}$$

$$l_t(t) \geq 0 \tag{8.51}$$

Equation (8.45) can be rewritten as

$$\dot{V}(t) \leq -\gamma_1 \frac{\beta}{2} \int_0^{l(t)} P(x,t)[w_x(x,t)]^2 dx$$

$$- \gamma_2 \frac{\beta \rho}{2} \int_0^{l(t)} [w_t(x,t) + l_t(t)w_x(x,t)]^2 dx$$

$$- \gamma_3 \frac{1}{2} z_1^2 - \gamma_4 \frac{m}{2} z_2^2 - \gamma_5 \frac{m}{2} z_3^2$$

$$+ \frac{1}{\gamma_\chi} (\xi N(\chi) - 1) \dot{\chi} + \varepsilon \tag{8.52}$$

where $\gamma_1 = \frac{\delta}{\beta P_{max}}$, $\gamma_2 = \frac{\alpha}{\beta} - \frac{1}{\rho \sigma_1}$, $\gamma_3 = 2\left(c_1 - \frac{1}{2}\beta c_1^2 P_{max} - \frac{1}{\sigma_3}\right)$, $\gamma_4 = \frac{2}{m}\left(c_2 - \frac{\beta P_{max}}{2}\right)$, $\gamma_5 = \frac{2}{m} c_3$, and $\varepsilon = \frac{1}{2c_l} \bar{d}_D^2 + \frac{1}{2}\left(\sigma_1 \beta + \frac{\alpha l(t)}{\sigma_2}\right) l(t) Q_{max}^2$.

We further obtain

$$\dot{V}(t) \leq -\lambda_1 [V_1(t) + V_b(t)] + \frac{1}{\gamma_\chi} (\xi N(\chi) - 1) \dot{\chi} + \varepsilon \tag{8.53}$$

where $\lambda_1 = \min(\gamma_1, \gamma_2, \gamma_3, \gamma_4, \gamma_5)$.

Combining (8.41) and (8.53), we have

$$\dot{V}(t) \leq -\lambda V(t) + \frac{1}{\gamma_\chi} (\xi N(\chi) - 1) \dot{\chi} + \varepsilon \tag{8.54}$$

where $\lambda = \lambda_1/\beta_3 > 0$.

Then multiplying Eq. (8.54) by $e^{\lambda t}$, and integrating of it, we have

$$V(t) \leq V(0)e^{-\lambda t} + \frac{\varepsilon}{\lambda}\left(1 - e^{-\lambda t}\right) \tag{8.55}$$

$$+ \frac{e^{-\lambda t}}{\gamma_\chi} \int_0^t (\xi N(\chi) - 1) \dot{\chi} e^{\lambda \tau} d\tau \tag{8.56}$$

Applying Lemma 2.8, we can conclude that $V(t)$, χ and $\int_0^t (\xi N(\chi) - 1) \dot{\chi} d\tau$ are bounded on $[0, t)$. This further implies that z_1, z_2, z_3, $w(x,t)$, $w_t(x,t)$ and $w_x(x,t)$ are all bounded. Note that

$$\left|\frac{\partial u_g(u_0)}{\partial u_0}\right| = \left|\frac{4}{\left(e^{u_0/u_M} + e^{-u_0/u_M}\right)^2}\right| \leq 1 \tag{8.57}$$

$$\left|\frac{\partial u_g(u_0)}{\partial u_0} u_0\right| = \left|\frac{4u_0}{\left(e^{u_0/u_M} + e^{-u_0/u_M}\right)^2}\right| \leq \frac{u_M}{2} \tag{8.58}$$

Then we can obtain that $\bar{\omega}$ is bounded from (8.57)–(8.58) and (8.29). This further implies that ω and u_0 are bounded.

According to Lemma 2.5, we obtain that $w(x, t)$ is uniformly bounded as follows:

$$w(x, t) \leq \sqrt{\frac{2l(t)}{\beta P_{\min}\beta_2} \left(V(0) + \frac{\varepsilon_0}{\lambda}\right)}$$

$$\leq \sqrt{\frac{2L}{\beta P_{\min}\beta_2} \left(V(0) + \frac{\varepsilon_0}{\lambda}\right)} \tag{8.59}$$

where $\varepsilon_0 = \varepsilon + \frac{\lambda}{\gamma_x} \int_0^t (\xi N(\chi) - 1)\dot{\chi}e^{-\lambda(t-\tau)}d\tau$.

Remark 8.3 In the above dynamic analysis, the process of controller design and the proof of the stability are based on the condition that $l_t(t) \geq 0$, that is extension of the flexible hose. When $l_t(t) < 0$, that is retraction of the flexible hose, the material derivative of a moving material should be redefined by $D(*)/Dt = \partial(*)/\partial t - l_t(t)\partial(*)/\partial x$. The system dynamic model, the control law and the system stability can be obtained by the similar procedure.

Remark 8.4 From the proof process, it is shown that the increase of the control gain c_2, c_3 and c_l will reduce the size of Ω and produce a better vibration reduction performance if the parameters c_1, α, β and $\sigma_i (i = 1, 2, 3)$ are chosen as proper values. Then we can conclude that the deflection of the hose $w(x, t)$ can be made arbitrarily small when the design control parameters are appropriately selected to satisfy the inequalities (8.46)–(8.51). However, very large control gains c_2, c_3 and c_l could lead to a high gain control problem. In practical applications, we should choose the parameters carefully on the condition that certain performance indicators can be satisfied.

Moreover, the proposed boundary control law (8.12) has more benefits than the existing control schemes [6]. The previous studies did not consider the varying length and the input constraint. The control scheme proposed in this chapter can handle input constraint effectively in the presence of varying length and varying speed. And the control design is based on the PDEs, which avoids the spillover problems associated with traditional truncated model-based approaches caused by ignoring high-frequency models in controller design.

8.5 Simulation

In this chapter, we use the finite difference method to simulate the system performance. By choosing the proper temporal and spatial step size to approximate the solution of the PDE system, the effectiveness of the proposed control law (8.12) is demonstrated by the finite difference method. The boundary disturbance $d(t)$ is given as $d(t) = 1.5 \sin(0.5t) + 1.5 \cos(0.5t)$, and the distributed disturbance is given

Table 8.1 Parameters of a flexible aerial refueling hose

Parameter	Description	Value
L	The total length of the link	16 m
ρ	The mass of the unit length	5.2 kg/m
D_h	The diameter of the hose	0.067 m
D_{drog}	The diameter of the drogue	0.61 m
m	The mass of the drogue	39.5 kg
g	Acceleration of gravity	9.8 m/s^2
ρ_{air}	The air density	1.29 kg/m^3
C_f	The skin friction coefficient	0.005
C_d	The pressure drag coefficient	0.45
C_{drog}	The drag coefficient	0.43

as $d_h(x, t) = 0.3 + 0.1 \sin(0.5xt) + 0.1 \sin(xt) + 0.1 \sin(1.5xt)$. The initial conditions are given as $w(x, 0) = 0.06x^2$ and $w_t(x, 0) = 0$. The constraint on the input u is given by $|u| \leq u_M = 6$. The parameters of the flexible hose are listed in Table 8.1.

For analyzing and verifying the control performance, the simulation results of the flexible hose with the proposed control laws in this study and with PD control laws are both under consideration. The PD control law is designed as follows:

$$u(t) = -k_p w(l(t), t) - k_d w_t(l(t), t) \tag{8.60}$$

with the parameters $k_p = 25$ and $k_d = 25$. And the proposed control (8.12) is chosen with the parameters $c = 2$, $c_1 = 25$, $c_2 = 15$, $c_3 = 15$ and $c_l = 30$. The length of the hose is $l(t) = l(0) + l_t(0)t + \frac{1}{2}l_{tt}(t)t^2$. The speed of the tanker is $v(t) = 100 + v_t(t)t$. The dynamic responses of the flexible hose system are simulated in the following four cases:

Case 1: $l_t(0) = 0.5$, $l_{tt}(t) = 0$, $v(t) = 0$
Case 2: $l_t(0) = 0.5$, $l_{tt}(t) = 0.05$, $v_t(t) = 0$
Case 3: $l_t(0) = 0.5$, $l_{tt}(t) = 0$, $v_t(t) = 0.1$
Case 4: $l_t(0) = 0.5$, $l_{tt}(t) = 0.05$, $v_t(t) = 0.1$

The simulation results of case 1, 2, 3 and 4 are shown in Figs. 8.2, 8.3, 8.4, 8.5, 8.6, 8.7, and 8.8, 8.9, respectively.

The dynamic responses without input control are shown in the Figs. 8.2, 8.4, 8.6 and 8.8 with dashed lines. It is clear that the vibration of the hose is large.

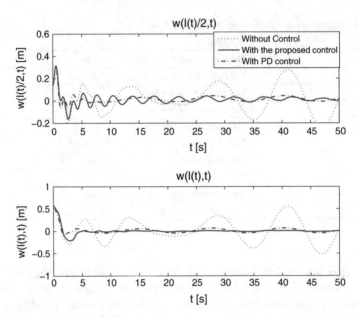

Fig. 8.2 Outputs of the flexible hose for case 1

Figure 8.2 shows the outputs $w(l(t))$ and $w(l(t)/2)$ of the hose with the proposed control (8.12) and the PD control (8.60) for case 1. We can see that the proposed control scheme and the PD control can both degrade the vibrations. Furthermore, with the proposed control law, the end point deflection $w(l(t), t)$ numerically converges to a small neighborhood of zero after 15 s, which means that the good performance of vibration suppressing can be obtained with the proposed control law. And the corresponding control inputs of the proposed control (8.12) and the PD control (8.60) for case 1 are shown in Fig. 8.3, which indicates the input value of the PD control law is larger than the input constraint. From the Figs. 8.4, 8.5, 8.6, 8.7, 8.8 and 8.9, we can see that both the proposed control law and the PD law can regulate the vibration greatly within 10 s when the acceleration of the hose ($l_{tt}(t)$) and the acceleration of the tanker ($v_t(t)$) are taken into account (for cases 2–4). Compared with case 1, the control effect with the proposed control method for these cases goes worse, but is still better than the PD control law.

From above analysis we can conclude that the effectiveness of the control scheme proposed in this chapter can be guaranteed in handing input constraint, even though the acceleration of the hose and the tanker, and the external disturbances are taken into account.

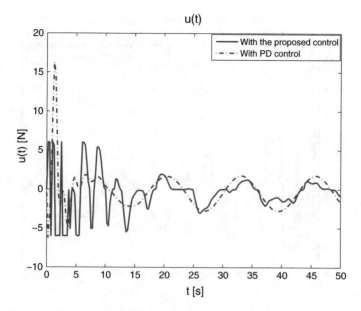

Fig. 8.3 The control input of the flexible hose for case 1

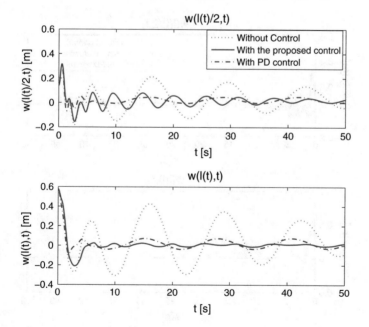

Fig. 8.4 Outputs of the flexible hose for case 2

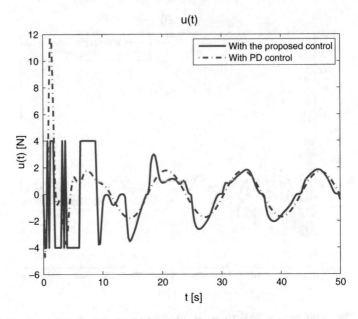

Fig. 8.5 The control input of the flexible hose for case 2

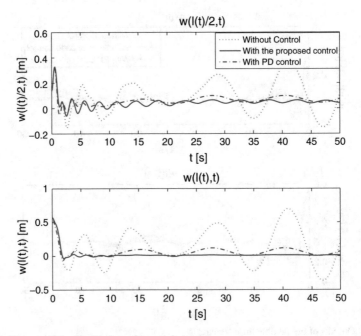

Fig. 8.6 Outputs of the flexible hose for case 3

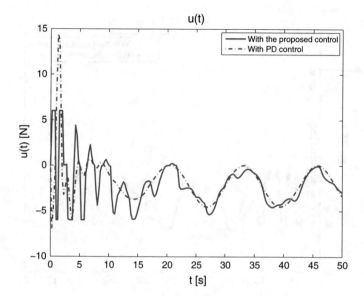

Fig. 8.7 The control input of the flexible hose for case 3

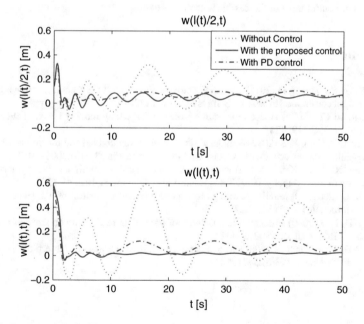

Fig. 8.8 Outputs of the flexible hose for case 4

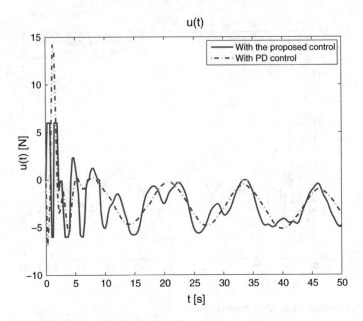

Fig. 8.9 The control input of the flexible hose for case 4

References

1. He W, Ge SS, Huang D (2015) Modeling and vibration control for a nonlinear moving string with output constraint. IEEE/ASME Trans Mech 1–12 (2015)
2. Li Y, Rahn CD (2000) Adaptive vibration isolation for axially moving beams. IEEE/ASME Trans Mech 5(4):419–428
3. Nguyen QC, Hong K-S (2012) Simultaneous control of longitudinal and transverse vibrations of an axially moving string with velocity tracking. J Sound Vib 331(13):3006–3019
4. Nguyen QC, Hong K-S (2012) Transverse vibration control of axially moving membranes by regulation of axial velocity. IEEE Trans Control Syst Technol 20(4):1124–1131
5. Zhihua Q (2002) An iterative learning algorithm for boundary control of a stretched moving string. Automatica 38(5):821–827
6. Zhu WD, Ni J (2000) Energetics and stability of translating media with an arbitrarily varying length. J Vib Acoust 122(3):295–304
7. Zhu WD, Ni J, Huang J (2001) Active control of translating media with arbitrarily varying length. J Vib Acoust 123(3):347–358

Chapter 9
Dynamic Modeling and Vibration Control for a Nonlinear Three-Dimensional Flexible Manipulator

9.1 Introduction

In the previous chapters, modeling and vibration control of the flexible mechanical systems are restricted to one dimensional space, and only transverse deformation is taken into account. However, flexible systems may move in a three-dimensional (3D) space in practical applications. The control performance will be affected if the coupling effects between motions in three directions are ignored. In spatial and industrial environment, flexible manipulators have been widely used due to their advantages such as light weight, fast motion and low energy consumption [3, 7]. For dynamic analysis, the flexible manipulator system is regarded as a distributed parameter system (DPS) which is mathematically represented by partial differential equations (PDEs) and ordinary differential equations (ODEs) [2, 5, 8], however, these works are only considered in one dimensional space. To improve accuracy and reliability analysis, modeling and control of the flexible manipulator system in a 3D space is necessary. Therefore, several works have been done in dynamics modeling and control design when the coupling effect are taken into account. In [1], design of boundary controllers actuated by hydraulic actuators at the top end is presented for global stabilization of a three-dimensional riser system. And design of boundary controllers implemented at the top end for global stabilization of a marine riser in a three dimensional space under environmental loadings based on PDEs is presented in [6]. To the best of our knowledge, there are few studies of modeling and control design for flexible manipulators in 3D space based on original PDEs, despite the significant progress of the control design for flexible systems in 3D space. The orientation and the vibrations of the manipulator should be considered simultaneously when a flexible manipulator moves in 3D space. In 3D space, there are strong couplings between deflections of a flexible manipulator along Y_b and Z_b axis, moreover, strong couplings between deflections and orientation, see Sect. 9.2. These couplings cause more difficulties to control a flexible manipulator in 3D space than the one in previous studies.

© Tsinghua University Press 2020
Z. Liu and J. Liu, *PDE Modeling and Boundary Control for Flexible Mechanical System*, Springer Tracts in Mechanical Engineering,
https://doi.org/10.1007/978-981-15-2596-4_9

In this chapter, we consider the trajectory tracking and the vibrations suppressing control problems of a flexible manipulator in 3D space. Based on the direct Lyapunov method, boundary control schemes with disturbance observers are proposed to regulate orientation and suppress elastic vibrations simultaneously. First, the model of a coupled three-dimensional flexible manipulator is derived by using Hamilton's principle and unit quaternion, and described by a set of PDEs and ODEs. Then, based on the proposed disturbance observers, boundary control laws are proposed to regulate orientation and suppress elastic vibration simultaneously. And, the closed-loop stability analysis avoids any simplification or discretization of the PDEs based on the direct Lyapunov method.

The rest of the chapter is organized as follows. The PDEs coupled with ODEs dynamics of a flexible manipulator in 3D space are derived in Sect. 9.2. Boundary control scheme and disturbance observers are designed and analyzed in Sect. 9.3. Numerical simulations are demonstrated in Sect. 9.4 to show the effectiveness of the proposed controller.

9.2   Problem Formulation

A typical flexible manipulator system in 3D space is shown in Fig. 9.1. The inertial frame is $OX_iY_iZ_i$, where O is the spherical driving joint of the flexible manipulator. The body fixed frame is $OX_bY_bZ_b$. Suppose the initial position of the beam is along the X_b axis. Suppose that the beam is inextensible, i.e., the deflections of the beam could occur in the OY_b and OZ_b directions only. The corresponding deflections with respect to $OX_bY_bZ_b$ are denoted by $y \equiv y(x, t)$ and $z \equiv z(x, t)$. Then, the position vector of the point P is

$$\mathbf{r} = \begin{bmatrix} x & y & z \end{bmatrix}^T \tag{9.1}$$

where x and t represent the independent spatial and time variable respectively. $\mathbf{Q}$ is the unit quaternion that describes the orientation of the body fixed frame $OX_bY_bZ_b$ with respect to the inertial frame $OX_iY_iZ_i$. $\boldsymbol{\omega} = \begin{bmatrix} \omega_1 & \omega_2 & \omega_3 \end{bmatrix}^T$ is the angular velocity of the body fixed frame with respect to the inertial frame, expressed in the body fixed frame. $\boldsymbol{\theta} = \begin{bmatrix} \theta_1 & \theta_2 & \theta_3 \end{bmatrix}^T$ is the Euler angle describing the orientation of the body fixed frame relative to the inertial frame $OX_iY_iZ_i$. The Euler angles in this paper are defined with the following sequence of rotations: a rotation about Z_b-axis by θ_3, then a rotation about Y_b-axis by θ_2, followed by a rotation about X_b-axis by θ_1.

To derive the model of this system, the expressions of kinetic energy E_k, potential energy E_p and virtual work δW should be introduced.

The kinetic energy of the flexible manipulator E_k can be represented as

$$E_k = \frac{1}{2}\boldsymbol{\omega}^T \mathbf{I}_h \boldsymbol{\omega} + \frac{1}{2}\rho \int_\Omega \left[\frac{d\mathbf{r}}{dt}\right]^T \frac{d\mathbf{r}}{dt} dx + \frac{1}{2}m_t \left[\frac{d\mathbf{r}(L, t)}{dt}\right]^T \frac{d\mathbf{r}(L, t)}{dt} \tag{9.2}$$

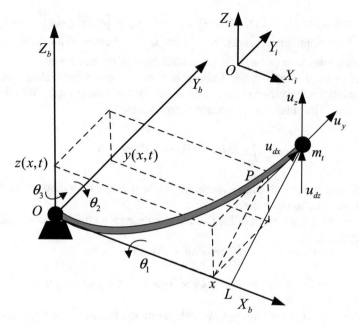

Fig. 9.1 Diagram of a three-dimensional flexible manipulator

where m is the point mass of tip payload, $\mathbf{I}_h$ is the inertia of the spherical driving joint, and $\mathbf{r} = \mathbf{D} + \mathbf{d}$. $\mathbf{D} = (x, 0, 0)^T$ denotes the position of the point P relative to the body fixed frame in undeformed state, and $\mathbf{d} = (0, y, z)^T$ represents the deformation of the point P relative to the body fixed frame. The time rate of change of the vector $\mathbf{r}$ relative to the inertial frame is denoted by $d\mathbf{r}/dt$, and that relative to the body fixed frame by $\dot{\mathbf{r}}$. Therefore, we have

$$\frac{d\mathbf{r}}{dt} = \dot{\mathbf{r}} + \boldsymbol{\omega} \times \mathbf{r} = \mathbf{d}_t + \boldsymbol{\omega} \times \mathbf{r} \tag{9.3}$$

The potential energy consists of elastic and gravitational energy. In this paper, gravitational energy is ignored. Therefore the potential energy due to deformation and the tension of the beam is given by

$$E_p = \frac{1}{2} E I \int_\Omega \left[\left(\frac{\partial^2 y}{\partial x^2} \right)^2 + \left(\frac{\partial^2 z}{\partial x^2} \right)^2 \right] dx + \frac{1}{2} T \int_\Omega \left[\left(\frac{\partial y}{\partial x} \right)^2 + \left(\frac{\partial z}{\partial x} \right)^2 \right] dx \tag{9.4}$$

where EI is the bending stiffness of the beam and T is the tension of the beam.

The virtual work done on the system is represented by

$$\delta W = (\boldsymbol{\tau} + \mathbf{d}_\tau)^T \delta \boldsymbol{\theta} + (\mathbf{u} + \mathbf{d}_u)^T \delta \mathbf{r}(L, t) \tag{9.5}$$

where $\mathbf{r}(L, t) = \begin{bmatrix} L & y(L, t) & z(L, t) \end{bmatrix}^T$. And $\mathbf{d}_\tau = \begin{bmatrix} d_{\tau x} & d_{\tau y} & d_{\tau z} \end{bmatrix}^T$ and $\mathbf{d}_u = \begin{bmatrix} d_{uy} & d_{uz} \end{bmatrix}^T$ are input disturbances. $\boldsymbol{\tau} = \begin{bmatrix} \tau_x & \tau_y & \tau_z \end{bmatrix}^T$ is the control torque at the spherical driving joint, $\mathbf{u} = \begin{bmatrix} u_y & u_z \end{bmatrix}^T$ is the control force at the end actuator.

Based on the assumption that the beam has small displacements, Hamilton's principle permits the derivation of equations of motion from energy quantities in a variational form. The Hamilton's principle is represented by

$$\int_{t_1}^{t_2} \left(\delta E_k - \delta E_p + \delta W \right) dt = 0 \tag{9.6}$$

where t_1 and t_2 are two time instants, $t_1 < t < t_2$ is the operation interval and δ denotes the variational operator.

The virtual displacement of the flexible manipulator relative to the inertial frame can be given by

$$\delta \mathbf{r} = \delta \boldsymbol{\theta} \times \mathbf{r} + \delta \mathbf{d} \tag{9.7}$$

$$\delta \left(d\mathbf{r}/dt \right) = \delta \boldsymbol{\omega} \times \mathbf{r} + \delta \boldsymbol{\theta} \times \left(\boldsymbol{\omega} \times \mathbf{r} + \mathbf{d}_t \right) + \left(d/dt \right) \left(\delta \mathbf{d} \right) \tag{9.8}$$

where $\delta \boldsymbol{\omega} = \left(d/dt \right) \left(\delta \boldsymbol{\theta} \right)$ denotes the virtual rotation rate of the body fixed frame.

Considering the variation of (9.2) and (9.4), we have

$$\int_{t_1}^{t_2} \delta E_k dt = \int_{t_1}^{t_2} \delta \boldsymbol{\omega}^T \left(\mathbf{I}_h \boldsymbol{\omega} \right) dt + \rho \int_{t_1}^{t_2} \int_{\Omega} \delta \left[\frac{d\mathbf{r}}{dt} \right]^T \frac{d\mathbf{r}}{dt} dx dt + m \int_{t_1}^{t_2} \delta \left[\frac{d\mathbf{r}(L, t)}{dt} \right]^T \frac{d\mathbf{r}(L, t)}{dt} dt$$

$$= - \int_{t_1}^{t_2} \int_{\Omega} \left(d/dt \right) \left(\mathbf{r} \times \mathbf{d}_t \right)^T \delta \boldsymbol{\theta} dx dt - \rho \int_{t_1}^{t_2} \int_{\Omega} \left(\mathbf{d}_t + \boldsymbol{\varpi} \right)^T \delta \mathbf{d} dx dt$$

$$- \int_{t_1}^{t_2} \left(\dot{\mathbf{I}}_s \boldsymbol{\omega} + \mathbf{I}_B \dot{\boldsymbol{\omega}} + \boldsymbol{\omega} \times \left(\mathbf{I}_B \boldsymbol{\omega} \right) \right)^T \delta \boldsymbol{\theta} dt - m \int_{t_1}^{t_2} \left(\mathbf{d}_{tt}(L, t) + \boldsymbol{\varpi}(L, t) \right)^T \delta \mathbf{d}(L, t) dt$$

$$- m \int_{t_1}^{t_2} \left(\mathbf{r}(L, t) \times \left(\mathbf{d}_{tt}(L, t) + \boldsymbol{\varpi}(L, t) \right) \right)^T \delta \boldsymbol{\theta} dt \tag{9.9}$$

where $\boldsymbol{\varpi} = 2\boldsymbol{\omega} \times \mathbf{d}_t + \dot{\boldsymbol{\omega}} \times \mathbf{r} + \boldsymbol{\omega} \times \left(\boldsymbol{\omega} \times \mathbf{r} \right)$, and $\mathbf{I}_B = \mathbf{I}_h + \mathbf{I}_s$, and

$$\mathbf{I}_s = \begin{bmatrix} I_{xx} & -I_{xy} & -I_{xz} \\ -I_{xy} & I_{yy} & -I_{yz} \\ -I_{xz} & -I_{yz} & I_{zz} \end{bmatrix} \tag{9.10}$$

where $I_{xx} = \rho \int_{\Omega} \left(y^2 + z^2 \right) dx$, $I_{yy} = \rho \int_{\Omega} \left(x^2 + z^2 \right) dx$, $I_{zz} = \rho \int_{\Omega} \left(x^2 + y^2 \right) dx$, $I_{xy} = \rho \int_{\Omega} xy dx$, $I_{xz} = \rho \int_{\Omega} xz dx$, $I_{yz} = \rho \int_{\Omega} yz dx$.

Considering Eq. (9.4), we can obtain

$$\delta E_p = EI \frac{\partial^2 y}{\partial x^2} \frac{\partial}{\partial x} \delta y|_0^L - EI \frac{\partial^3 y}{\partial x^3} \delta y|_0^L + EI \int_{\Omega} \frac{\partial^4 y}{\partial x^4} \delta y dx$$

$$+ EI \frac{\partial^2 z}{\partial x^2} \frac{\partial}{\partial x} \delta z|_0^L - EI \frac{\partial^3 z}{\partial x^3} \delta z|_0^L + EI \int_{\Omega} \frac{\partial^4 z}{\partial x^4} \delta z dx$$

$$+ T \frac{\partial y}{\partial x} \delta y |_0^L - T \int_\Omega \frac{\partial^2 y}{\partial x^2} \delta y dx + T \frac{\partial z}{\partial x} \delta z |_0^L - T \int_\Omega \frac{\partial^2 z}{\partial x^2} \delta z dx \qquad (9.11)$$

Combining with (9.7), (9.5) can be rewritten as

$$\begin{aligned}
\delta W &= (\boldsymbol{\tau} + \mathbf{d}_\tau)^T \delta\boldsymbol{\theta} + (\mathbf{u} + \mathbf{d}_u)^T \delta \mathbf{r}(L, t) \\
&= (\boldsymbol{\tau} + \mathbf{d}_\tau)^T \delta\boldsymbol{\theta} + (\mathbf{u} + \mathbf{d}_u)^T \delta \mathbf{d} + (\mathbf{u} + \mathbf{d}_u)^T (\delta\boldsymbol{\theta} \times \mathbf{r}(L, t)) \\
&= (\boldsymbol{\tau} + \mathbf{d}_\tau)^T \delta\boldsymbol{\theta} + (\mathbf{u} + \mathbf{d}_u)^T \delta \mathbf{d} + (\mathbf{r}(L, t) \times (\mathbf{u} + \mathbf{d}_u))^T \delta\boldsymbol{\theta} \qquad (9.12)
\end{aligned}$$

Noting that the variations $\delta\theta_1$, $\delta\theta_2$, $\delta\theta_3$, δy and δz are arbitrary, and using unit-quaternions to represent the orientation, we obtain the coupled governing equations of the system as

$$\dot{\eta} = -\frac{1}{2}\mathbf{q}^T \boldsymbol{\omega} \qquad (9.13)$$

$$\dot{\mathbf{q}} = \frac{1}{2}\left(\mathbf{q}^\times + \eta \mathbf{I}_3\right)\boldsymbol{\omega} \qquad (9.14)$$

$$\mathbf{I}_h \dot{\boldsymbol{\omega}} + \boldsymbol{\omega} \times (\mathbf{I}_h \boldsymbol{\omega}) + M + N = \boldsymbol{\tau} + \mathbf{d}_\tau \qquad (9.15)$$

$$\rho \begin{bmatrix} y_{tt} \\ z_{tt} \end{bmatrix} + \rho\phi\varpi + EI \begin{bmatrix} y_{xxxx} \\ z_{xxxx} \end{bmatrix} - T \begin{bmatrix} y_{xx} \\ z_{xx} \end{bmatrix} = 0 \qquad (9.16)$$

and the boundary conditions of the systems as

$$m \begin{bmatrix} y_{tt}(L, t) \\ z_{tt}(L, t) \end{bmatrix} + m\phi\varpi(L, t) - EI \begin{bmatrix} y_{xxx}(L, t) \\ z_{xxx}(L, t) \end{bmatrix} + T \begin{bmatrix} y_x(L, t) \\ z_x(L, t) \end{bmatrix} = \begin{bmatrix} u_y \\ u_z \end{bmatrix} + \begin{bmatrix} d_{uy} \\ d_{uz} \end{bmatrix} \qquad (9.17)$$

$$y_{xx}(L, t) = z_{xx}(L, t) = 0 \qquad (9.18)$$

$$y_x(0, t) = z_x(0, t) = y(0, t) = z(0, t) = 0 \qquad (9.19)$$

where M, N, ϕ and ϖ are defined as follows

$$M = EI \begin{bmatrix} 0 & z_{xx}(0, t) & -y_{xx}(0, t) \end{bmatrix}^T \qquad (9.20)$$

$$N = T \begin{bmatrix} 0 & z(L, t) & -y(L, t) \end{bmatrix}^T \qquad (9.21)$$

$$\phi = \begin{bmatrix} 0 & 1 & 0 \\ 0 & 0 & 1 \end{bmatrix} \qquad (9.22)$$

$$\varpi = 2 \begin{bmatrix} 0 & -\omega_3 & \omega_2 \\ \omega_3 & 0 & -\omega_1 \\ -\omega_2 & \omega_1 & 0 \end{bmatrix} \begin{bmatrix} 0 \\ y_t \\ z_t \end{bmatrix} + \begin{bmatrix} 0 \\ \omega_1\omega_2 + \dot{\omega}_2 \\ \omega_1\omega_3 - \dot{\omega}_3 \end{bmatrix} x$$

$$+ \begin{bmatrix} -(\omega_2^2 + \omega_3^2) & \omega_1\omega_2 - \dot{\omega}_3 & \omega_1\omega_3 + \dot{\omega}_2 \\ \omega_1\omega_2 + \dot{\omega}_3 & -(\omega_1^2 + \omega_3^2) & \omega_2\omega_3 - \dot{\omega}_1 \\ \omega_1\omega_3 - \dot{\omega}_2 & \omega_2\omega_3 + \dot{\omega}_1 & -(\omega_1^2 + \omega_2^2) \end{bmatrix} \begin{bmatrix} 0 \\ y \\ z \end{bmatrix} \qquad (9.23)$$

and $\mathsf{Q} = [\eta \ \mathsf{q}^T]^T$ is a unit quaternion, which is composed of a vector component $\mathsf{q} = [q_1 \ q_2 \ q_3]^T$ and a scalar component η, satisfying

$$\eta^2 + \mathsf{q}^T \mathsf{q} = 1,$$

and $\mathsf{q}^\times$ is the skew-symmetric matrix operator and is defined by

$$\mathsf{q}^\times = \begin{bmatrix} 0 & -q_3 & q_2 \\ q_3 & 0 & -q_1 \\ -q_2 & q_1 & 0 \end{bmatrix} \qquad (9.24)$$

The function that maps Euler angles to their corresponding unit quaternion is

$$\mathsf{Q} = \begin{bmatrix} c_{\theta_1/2} \, c_{\theta_2/2} \, c_{\theta_3/2} + s_{\theta_1/2} \, s_{\theta_2/2} \, s_{\theta_3/2} \\ s_{\theta_1/2} \, c_{\theta_2/2} \, c_{\theta_3/2} - c_{\theta_1/2} \, s_{\theta_2/2} \, s_{\theta_3/2} \\ c_{\theta_1/2} \, s_{\theta_2/2} \, c_{\theta_3/2} + s_{\theta_1/2} \, c_{\theta_2/2} \, s_{\theta_3/2} \\ c_{\theta_1/2} \, c_{\theta_2/2} \, s_{\theta_3/2} - s_{\theta_1/2} \, s_{\theta_2/2} \, c_{\theta_3/2} \end{bmatrix} \qquad (9.25)$$

where $c_\theta = \cos(\theta)$ and $s_\theta = \sin(\theta)$.

Remark 9.1 The dynamics of the flexible manipulator are derived by using extended Hamilton's principle [2, 4, 6]. The dynamics consisting of the rigid body (driving joint) and a flexible beam rigidly attached to it, therefore, is given by the set of ODEs and the set of the PDEs along with the boundary conditions. It is clearly seen that the two sets of equations are strongly coupled.

In case the displacements of the beam are zero, the resulting form of (9.13)–(9.15) is the well-known attitude dynamics of a rigid body. Also in absence of the dynamics of the rigid body (driving joint), these equations reduce to the usual equations for a three dimensional marine riser [6] or an Euler-Bernoulli beam [4].

Assumption 9.1 For the time derivative of boundary disturbances $\dot{d}_{\tau x}, \dot{d}_{\tau y}, \dot{d}_{\tau z}, \dot{d}_{uy}$ and $\dot{d}_{uz}$, we assume that there exist positive constants $\bar{d}_{\tau x}, \bar{d}_{\tau y}, \bar{d}_{\tau z}, \bar{d}_{uy}$ and $\bar{d}_{uz}$, such that $|\dot{d}_{\tau x}| \le \bar{d}_{\tau x}, |\dot{d}_{\tau y}| \le \bar{d}_{\tau y}, |\dot{d}_{\tau z}| \le \bar{d}_{\tau z}, |\dot{d}_{uy}| \le \bar{d}_{uy}, |\dot{d}_{uz}| \le \bar{d}_{uz}, \forall t \in [0, \infty)$.

9.3   Control Design

In this paper, the control aim is to design a controller to regulate orientation and suppress elastic vibration simultaneously in the presence of external disturbances, i.e., from any condition to desired orientation $\mathbf{Q}_d=[\eta_d \ \mathbf{q}_d^T]^T$ and desired angular velocity ω_d, and the deflections tend to a small neighborhood of zero. First we define the tracking error $\tilde{\mathbf{Q}} = [\tilde{\eta} \ \tilde{\mathbf{q}}^T]^T$ and $\tilde{\mathbf{q}} = \begin{bmatrix} \tilde{q}_1 & \tilde{q}_2 & \tilde{q}_3 \end{bmatrix}^T$, where $\tilde{\eta} = \eta\eta_d + \mathbf{q}_d^T\mathbf{q}$ and $\tilde{\mathbf{q}} = \eta_d\mathbf{q} - \eta\mathbf{q}_d + \mathbf{q}^\times\mathbf{q}_d$.

Then we have

$$\tilde{\eta}^2 + \tilde{\mathbf{q}}^T\tilde{\mathbf{q}} = 1 \tag{9.26}$$

$$\dot{\tilde{\eta}} = -\frac{1}{2}\tilde{\mathbf{q}}^T\tilde{\omega} \tag{9.27}$$

$$\dot{\tilde{\mathbf{q}}} = G\tilde{\omega} \tag{9.28}$$

where

$$G = \frac{1}{2}\left(\tilde{\mathbf{q}}^\times + \tilde{\eta}\mathbf{I}_3\right) \tag{9.29}$$

$$\tilde{\omega} = \omega - \mathbf{R}\left(\tilde{\mathbf{Q}}\right)\omega_d \tag{9.30}$$

$$\mathbf{R}\left(\tilde{\mathbf{Q}}\right) = \left(\tilde{\eta}^2 - \tilde{\mathbf{q}}^T\tilde{\mathbf{q}}\right)\mathbf{I}_3 + 2\tilde{\mathbf{q}}\tilde{\mathbf{q}}^T - 2\tilde{\eta}\tilde{\mathbf{q}}^\times \tag{9.31}$$

Suppose $\omega_d = \mathbf{0}$, we have $\tilde{\omega} = \omega$. Then we need to find inputs τ and $\mathbf{u}$ such that for all initial conditions the states of the closed-loop system can be stabilized.

Considering the system dynamics described by (9.13)–(9.19) with Assumption 9.1, the boundary control laws τ and $\mathbf{u}$ are designed by

$$\tau = -G^T\mathbf{k}_p\tilde{\mathbf{q}} - \mathbf{k}_d\tilde{\omega} - k_f G^T G\tilde{\omega} - \mathbf{k}_q\tilde{\mathbf{q}} - \hat{\mathbf{d}}_\tau \tag{9.32}$$

$$\begin{aligned}
\mathbf{u} = &- \mathbf{k}_u\mathbf{u}_0 - EI\phi\mathbf{d}_{xxx}(L,t) + T\phi\mathbf{d}_x(L,t) - \hat{\mathbf{d}}_u \\
&- m\phi\left[k_1\mathbf{d}_x(L,t) + k_1\tilde{\omega}\times\mathbf{d}_x(L,t) - k_2\mathbf{d}_{xxx}(L,t) - k_2\tilde{\omega}\times\mathbf{d}_{xxx}(L,t)\right]
\end{aligned} \tag{9.33}$$

with disturbance observers

$$\begin{aligned}
\dot{\mathbf{z}}_1 &= -K_1\left(\tau - M - N\right) - K_1\hat{\mathbf{d}}_\tau \\
\hat{\mathbf{d}}_\tau &= \mathbf{z}_1 + K_1\mathbf{I}_\mathbf{h}\omega
\end{aligned} \tag{9.34}$$

and

$$\dot{z}_2 = -K_2 \left(\mathbf{u} + EI\phi \mathbf{d}_{xxx}(L, t) - T\phi \mathbf{d}_x(L, t) \right) - K_2 \hat{\mathbf{d}}_u$$

$$\hat{\mathbf{d}}_u = \mathbf{z}_2 + K_2 m\phi \left[\frac{d\mathbf{r}(L, t)}{dt} \right] \tag{9.35}$$

where $\mathbf{k}_p = \text{diag}\left(k_{p1}, k_{p2}, k_{p3}\right), \mathbf{k}_d = \text{diag}\left(k_{d1}, k_{d2}, k_{d3}\right), \mathbf{k}_q = \text{diag}\left(k_{q1}, k_{q2}, k_{q3}\right),$ $\mathbf{k}_u = \text{diag}\left(k_{u1}, k_{u2}\right), K_1 = \text{diag}\left(K_{11}, K_{12}, K_{13}\right)$ and $K_2 = \text{diag}\left(K_{21}, K_{22}\right)$ are positive definite symmetric matrices, k_1, k_2, k_f are positive constants, and the auxiliary term is defined as

$$\mathbf{u}_0 = \phi \left[\frac{d\tilde{\mathbf{r}}_L}{dt} + k_1 \mathbf{d}_x(L, t) - k_2 \mathbf{d}_{xxx}(L, t) \right] \tag{9.36}$$

Remark 9.2 The control laws (9.32) and (9.33) are designed based on the distributed parameter systems and the the spillover problems associated with traditional truncated model-based approaches by ignoring high-frequency modes are avoided.

Theorem 9.1 *Suppose the system (9.13)–(9.19) satisfies Assumption 9.1. Using the proposed control scheme (9.32) and (9.33) with the disturbance observers (9.34) and (9.35), then the closed-loop system is:*

(i) uniform bounded (UB): the state of the system $y(x, t)$, $z(x, t)$ and $\tilde{q}_i$ $(i = 1, 2, 3)$ will remain in the compact sets

$$\Omega_1 = \{y(x, t), z(x, t) \in R \,||y(x, t)|, |z(x, t)| \le D_1\} \tag{9.37}$$

$$\Omega_2 = \{\tilde{q}_i \in R \,||\tilde{q}_i| \le D_2, i = 1, 2, 3\} \tag{9.38}$$

where the constants $D_1 = \sqrt{\frac{2L}{\beta k_2 T \alpha_2}} \left(V(0) + \frac{\varepsilon}{\lambda}\right), D_2 = \sqrt{\frac{2}{\beta k_2 \alpha_2 \lambda_{\min}(\mathbf{k}_p)}} \left(V(0) + \frac{\varepsilon}{\lambda}\right).$

(ii) uniform ultimate bounded (UUB): the state of the system $y(x, t)$, $z(x, t)$ and $\tilde{q}_i$ $(i = 1, 2, 3)$ will converge to the compact sets

$$\Omega_3 = \left\{ y(x, t), z(x, t) \in R \,\middle|\, \lim_{t \to \infty} |y(x, t)|, \lim_{t \to \infty} |z(x, t)| \le D_3 \right\} \tag{9.39}$$

$$\Omega_4 = \left\{ \tilde{q}_i \in R \,\middle|\, \lim_{t \to \infty} |\tilde{q}_i| \le D_4, i = 1, 2, 3 \right\} \tag{9.40}$$

where the constants $D_3 = \sqrt{\frac{2L\varepsilon}{\beta k_2 T \alpha_2 \lambda}}, D_4 = \sqrt{\frac{2\varepsilon}{\beta k_2 \alpha_2 \lambda_{\min}(\mathbf{k}_p)\lambda}}.$ And $\lambda_{\min}(\mathbf{k}_p)$ denotes the minimum eigenvalues of $\mathbf{k}_p$.

Proof The proof of Theorem can be found in Appendix.

Remark 9.3 The proof process is the control law design process based on the Lyapunov's direct method. When the parameters α, β, k_1, k_2, k_f and $\mathbf{k}_p$ are chosen as proper values, it is shown that the increase of the control gain matrices k_q, k_d and k_u will bring about a larger λ_1, which will result in a larger λ. Then the increscent value of λ will reduce the size of Ω_3 and Ω_4, i.e., produce a better vibration reduction performance and orientation tracking performance. Then we can conclude that the deflection of the flexible manipulator $y(x, t)$ and $z(x, t)$, and the orientation tracking errors $\tilde{q}_i$ ($i = 1, 2, 3$) can be made arbitrarily small when the design control parameters are appropriately selected to satisfy the inequalities (9.83)–(9.95). However, a very large control gains design could lead to a high gain control problem. In practical applications, we should choose the control gain matrices as small as possible on the condition that certain performance indicators can be satisfied.

Remark 9.4 For the system (9.13)–(9.19), if there is no external disturbances or the disturbances are invariable, the exponential stability of the closed-loop system can be obtained using the proposed control scheme (9.32) and (9.33) with the disturbance observers (9.34) and (9.35). That is, the desired orientations are achieved

$$\lim_{t \to \infty} \tilde{\mathbf{Q}} = 0, \lim_{t \to \infty} \tilde{\omega} = \lim_{t \to \infty} \omega = 0, \tag{9.41}$$

and the deflections of the flexible beam are suppressed, i.e.,

$$\lim_{t \to \infty} y(x, t) = \lim_{t \to \infty} z(x, t) = 0 \tag{9.42}$$

9.4 Simulation

The effectiveness of the proposed boundary control laws (9.32) and (9.33) with disturbance observers (9.34) and (9.35) is illustrated by numerical simulations. The desired Euler angles are $\theta_{1d} = 0$, $\theta_{2d} = \frac{\pi}{3}$, $\theta_{3d} = \frac{\pi}{3}$. According to (9.25), the desired unit quaternion $\mathbf{Q}_d = [\eta_d \ \mathbf{q}_d^T]^T$ can be obtained. The external disturbances are given as $d_{uy} = d_{uz} = 0.2 + 0.1 \sin(0.5t) + 0.2 \sin(t)$, $d_{\tau y} = d_{\tau z} = (1 + \sin(0.1t) + 3 \sin(0.5t) + 5 \sin(t))/5$ and $d_{\tau x} = (1 + 0.5 \sin(0.1t) + 1 \sin(t))/5$. The initial conditions are given as $y(x, 0) = z(x, 0) = 0.25x/L$ and $y_t(x, 0) = z_t(x, 0) = 0$. The parameters of the flexible manipulator are listed in Table 9.1.

For analyzing and verifying the control performance, the dynamic responses of the system are simulated in the following two cases:

Case 1: Without control input: $\tau = 0$ and $\mathbf{u} = 0$.

Case 2: Applying the proposed control (9.32) and (9.33) and disturbance observers (9.34) and (9.35) with parameters $\mathbf{k}_p = \text{diag}(15, 10, 15)$, $\mathbf{k}_q = \text{diag}(15, 10, 15)$, $\mathbf{k}_d = \text{diag}(15, 10, 15)$, $\mathbf{k}_u = \text{diag}(40, 40)$, $k_1 = 2$, $k_2 = 2$, $k_f = 15$, $K_1 = \text{diag}(40, 40, 40)$ and $K_2 = \text{diag}(20, 20)$.

The dynamic responses without input control are shown in Figs. 9.2, 9.3 and 9.4. It is clear that the vibration amplitude of the flexible manipulator in Y_b and Z_b directions are large. The simulation results of case 2 are shown in Figs. 9.5, 9.6, 9.7,

Table 9.1 Parameters of a three-dimensional flexible manipulator

Parameter	Description	Value
L	The length of the flexible beam	1 m
EI	The bending stiffness of the beam	8 Nm^2
T	The tension of the beam	5 N
m	The point mass tip payload	2.0 kg
I_{h1}	The inertia of the spherical driving joint	0.15 kgm^2
I_{h2}	The inertia of the spherical driving joint	0.2 kgm^2
I_{h3}	The inertia of the spherical driving joint	0.2 kgm^2
ρ	The mass of the unit length	0.2 kg/m

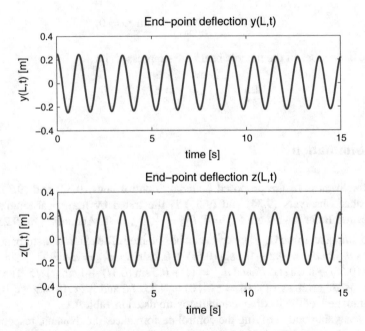

Fig. 9.2 Displacement of the beam's endpoint at Y and Z directions without control

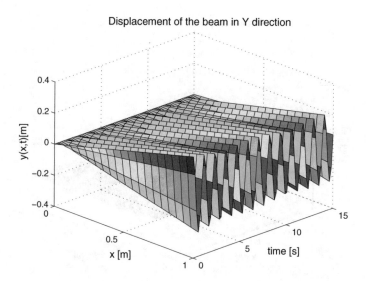

Fig. 9.3 Displacement of the beam at Y direction without control

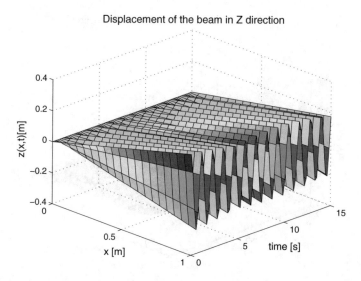

Fig. 9.4 Displacement of the beam at Z direction without control

Displacement of the beam in Y direction

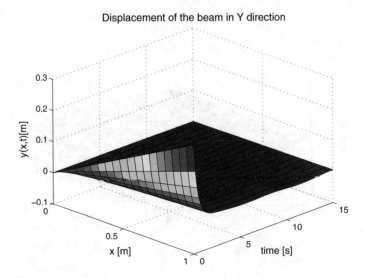

Fig. 9.5 Displacement of the beam at Y direction with boundary control

Displacement of the beam in Z direction

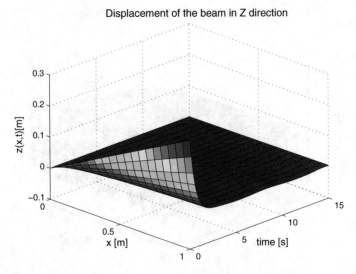

Fig. 9.6 Displacement of the beam at Z directions with boundary control

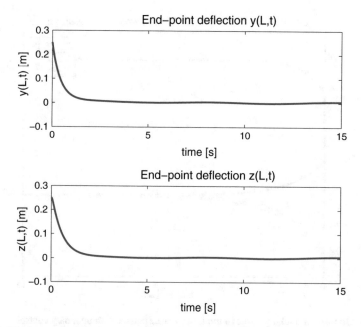

Fig. 9.7 Displacement of the beam's endpoint at Y and Z directions with boundary control

9.8, 9.9, 9.10, 9.11, 9.12 and 9.13. Figures 9.5 and 9.6 show the deflections in Y_b and Z_b directions. The tail end vibrations $y(L, t)$ and $z(L, t)$ are shown in Fig. 9.7. Figure 9.8 gives the orientation tracking errors of the flexible manipulator. And the angular velocity tracking errors are given in Fig. 9.9. From Figs. 9.5, 9.6, 9.7, 9.8 and 9.9, we can see that the proposed control scheme (9.32) and (9.33) can regulate the desired orientation and suppress the vibrations greatly within 5 s, and the orientation tracking errors and the vibrations can converge to a small neighborhood of zero after 5 s, which means that the good performance can be obtained with the proposed control law.

Moreover, the control inputs are shown in Figs. 9.10 and 9.11, and the disturbances and their estimates are shown in Figs. 9.12 and 9.13. The Figs. 9.12 and 9.13 illustrate that the estimations of input disturbance can be fast and accurately converge to their true values. Using the disturbance observer rather than a robust control method to solve the problem of disturbances is advantageous to the safe use of the actuators and the attenuation of vibration of the controlled plant.

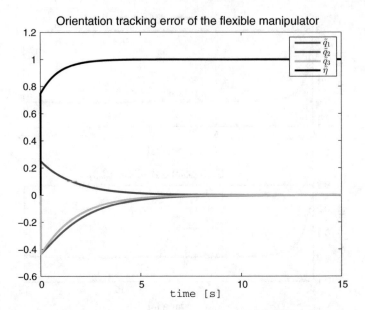

Fig. 9.8 Orientation tracking errors of the flexible manipulator with boundary control

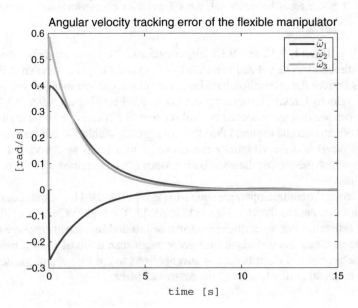

Fig. 9.9 Velocity tracking errors of the flexible manipulator with boundary control

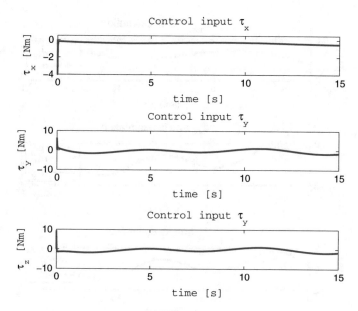

Fig. 9.10 The control torque at the spherical driving joint

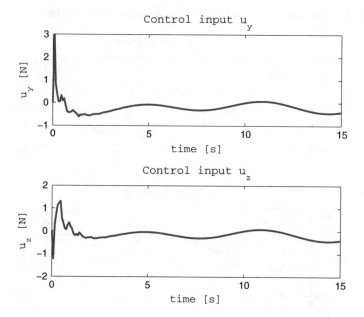

Fig. 9.11 The control force at the end actuator

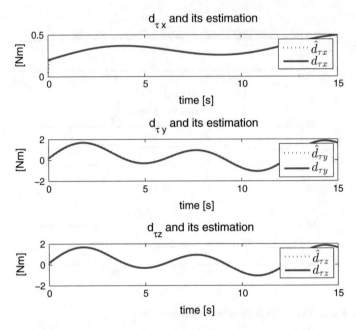

Fig. 9.12 The input disturbances and their estimates at the spherical driving joint

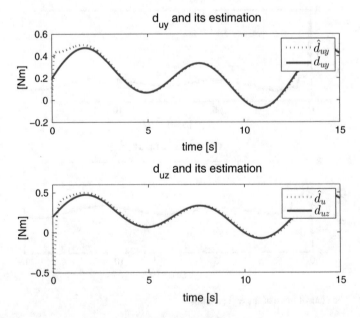

Fig. 9.13 The input disturbances and their estimates at the end actuator

Appendix 1: Proof of the Theorem

Consider the Lyapunov candidate function as

$$V = V_1 + V_2 + V_3 + V_4 + V_o \tag{9.43}$$

where

$$V_1 = \frac{1}{2}\beta k_2 \tilde{\boldsymbol{\omega}}^T \mathbf{I}_h \tilde{\boldsymbol{\omega}} + \frac{1}{2}\beta k_2 \tilde{\mathbf{q}}^T \mathbf{k}_p \tilde{\mathbf{q}} + \frac{1}{2}\beta k_2 \rho \int_\Omega \left[\phi \frac{d\tilde{\mathbf{r}}}{dt}\right]^T \left[\phi \frac{d\tilde{\mathbf{r}}}{dt}\right] dx \tag{9.44}$$

$$V_2 = \frac{1}{2}\beta k_2 EI \int_\Omega \left[(y_{xx})^2 + (z_{xx})^2\right] dx + \frac{1}{2}\beta k_2 T \int_\Omega \left[(y_x)^2 + (z_x)^2\right] dx \tag{9.45}$$

$$V_3 = \frac{1}{2}m\mathbf{u}_0^T \mathbf{u}_0 \tag{9.46}$$

$$V_4 = \alpha\rho \int_\Omega x \left[\phi \frac{d\tilde{\mathbf{r}}}{dt}\right]^T [\phi \mathbf{d}_x] dx + \alpha\tilde{\mathbf{q}}^T \mathbf{I}_h \tilde{\boldsymbol{\omega}} + \alpha\rho \int_\Omega x \left[\phi_1 \frac{d\tilde{\mathbf{r}}}{dt}\right]^T [\phi_2 \tilde{\mathbf{q}}] dx \tag{9.47}$$

$$V_o = \frac{1}{2}\tilde{\mathbf{d}}_\tau^T \tilde{\mathbf{d}}_\tau + \frac{1}{2}\tilde{\mathbf{d}}_u^T \tilde{\mathbf{d}}_u \tag{9.48}$$

where the estimate errors are $\tilde{\mathbf{d}}_\tau = \mathbf{d}_\tau - \hat{\mathbf{d}}_\tau$ and $\tilde{\mathbf{d}}_u = \mathbf{d}_u - \hat{\mathbf{d}}_u$, and $\tilde{\mathbf{r}} = \mathbf{r}$, $d\tilde{\mathbf{r}}/dt = \mathbf{d}_t + \tilde{\boldsymbol{\omega}} \times \tilde{\mathbf{r}}$, and $\phi_1 = \begin{bmatrix} 0 & 1 & 0 \\ 0 & 0 & -1 \end{bmatrix}$, $\phi_2 = \begin{bmatrix} 0 & 0 & 1 \\ 0 & 1 & 0 \end{bmatrix}$.

To motivate the followings, we first focus our attention on Eq. (9.47) and we can obtain

$$|V_4| \leq \alpha\rho L \int_\Omega \left[\phi \frac{d\tilde{\mathbf{r}}}{dt}\right]^T \left[\phi \frac{d\tilde{\mathbf{r}}}{dt}\right] dx + \frac{1}{2}\alpha\rho L \int_\Omega [\phi \mathbf{d}_x]^T [\phi \mathbf{d}_x] dx$$
$$+ \frac{1}{2}\alpha\rho L^2 [\phi_2 \tilde{\mathbf{q}}]^T [\phi_2 \tilde{\mathbf{q}}] + \frac{1}{2}\alpha\tilde{\mathbf{q}}^T \mathbf{I}_h \tilde{\mathbf{q}} + \frac{1}{2}\alpha\tilde{\boldsymbol{\omega}}^T \mathbf{I}_h \tilde{\boldsymbol{\omega}}$$
$$\leq \alpha_1 (V_1 + V_2) \tag{9.49}$$

where $\alpha_1 = \max\left[\frac{2\alpha\rho L}{k_2 \beta\rho}, \frac{\alpha\rho L}{k_2 \beta T}, \frac{\alpha I_{h1}}{k_2 \beta k_{p1}}, \frac{\alpha\rho L^2 + \alpha I_{hi}}{k_2 \beta k_{pi}}, \frac{\alpha}{k_2 \beta}\right], i = 2, 3$.

Then, we have

$$-\alpha_1 (V_1 + V_2) \leq V_4 \leq \alpha_1 (V_1 + V_2)$$

Choosing β, k_2 and $\mathbf{k}_p$ to satisfy the inequality $0 < \alpha_1 < 1$ yields $\alpha_2 = 1 - \alpha_1 > 0$, $\alpha_3 = 1 + \alpha_1 > 1$.

Then we further have

$$0 \leq \alpha_2 (V_1 + V_2) \leq V_1 + V_2 + V_4 \leq \alpha_3 (V_1 + V_2)$$

Given the Lyapunov function in (9.43), we obtain

$$0 \le \alpha_2 \left(V_1 + V_2 + V_3 + V_o\right) \le V \le \alpha_3 \left(V_1 + V_2 + V_3 + V_o\right) \tag{9.50}$$

Differentiating (9.43) with respect to time, we have

$$\dot{V} = \dot{V}_1 + \dot{V}_2 + \dot{V}_3 + \dot{V}_4 + \dot{V}_o \tag{9.51}$$

The term $\dot{V}_1$ is rewritten as

$$\dot{V}_1 = \dot{V}_{11} + \dot{V}_{12} + \dot{V}_{13} \tag{9.52}$$

where

$$\dot{V}_{11} = k_2 \beta \tilde{\omega}^T \mathbf{I}_h \dot{\tilde{\omega}} + k_2 \beta \tilde{\omega}^T \left(\tilde{\omega} \times (\mathbf{I}_h \tilde{\omega})\right) \tag{9.53}$$

$$\dot{V}_{12} = k_2 \beta \rho \int_\Omega \left[\phi \frac{d\tilde{\mathbf{r}}}{dt}\right]^T \left[\phi \frac{d^2\tilde{\mathbf{r}}}{dt^2}\right] dx \tag{9.54}$$

$$\dot{V}_{13} = k_2 \beta \tilde{\mathbf{q}}^T \mathbf{k}_p \dot{\tilde{\mathbf{q}}} \tag{9.55}$$

Considering (9.15) and (9.30), we have

$$\dot{V}_{11} = k_2 \beta \tilde{\omega}^T \left(\boldsymbol{\tau} + \mathbf{d}_\tau - M - N\right) \tag{9.56}$$

and

$$
\begin{aligned}
\dot{V}_{12} &= k_2 \beta \rho \int_\Omega \left[\phi \frac{d\tilde{\mathbf{r}}}{dt}\right]^T \left[\phi \frac{d^2\tilde{\mathbf{r}}}{dt^2}\right] dx \\
&= k_2 \beta \rho \int_\Omega [\phi \mathbf{d}_t]^T \left[\phi \left(\tilde{\omega} + \mathbf{d}_{tt}\right)\right] dx + k_2 \beta \rho \int_\Omega \left[\phi \left(\tilde{\omega} \times \tilde{\mathbf{r}}\right)\right]^T \left[\phi \left(\tilde{\omega} + \mathbf{d}_{tt}\right)\right] dx \\
&= k_2 \beta \rho \int_\Omega [\phi \mathbf{d}_t]^T \left[\phi \left(\tilde{\omega} + \mathbf{d}_{tt}\right)\right] dx + k_2 \beta \rho \int_\Omega \tilde{\omega}^T \left(\tilde{\mathbf{r}} \times \left[\phi^T \phi \left(\tilde{\omega} + \mathbf{d}_{tt}\right)\right]\right) dx
\end{aligned}
\tag{9.57}
$$

Since

$$
\begin{aligned}
\rho \int_\Omega [\phi \mathbf{d}_t]^T \left[\phi \left(\tilde{\omega} + \mathbf{d}_{tt}\right)\right] dx = &-\int_\Omega y_t \left(EI y_{xxxx} - T y_{xx}\right) dx \\
&- \int_\Omega z_t \left(EI z_{xxxx} - T z_{xx}\right) dx
\end{aligned}
\tag{9.58}
$$

and

$$\int_\Omega \rho \tilde{\mathbf{r}} \times \left[\boldsymbol{\phi}^T \boldsymbol{\phi} \left(\tilde{\boldsymbol{\omega}} + \mathbf{d}_{tt} \right) \right] dx = T\tilde{\mathbf{r}} \times \mathbf{d}_x(L, t) - EI\tilde{\mathbf{r}} \times \mathbf{d}_{xxx}(L, t) + M + N$$
(9.59)

we have

$$\dot{V}_{12} = k_2 \beta \tilde{\boldsymbol{\omega}}^T \left(M + N - EI\tilde{\mathbf{r}} \times \mathbf{d}_{xxx}(L, t) + T\tilde{\mathbf{r}} \times \mathbf{d}_x(L, t) \right)$$
$$- k_2 \beta \int_\Omega z_t \left(EI z_{xxxx} - T z_{xx} \right) dx - k_2 \beta \int_\Omega y_t \left(EI y_{xxxx} - T y_{xx} \right) dx \quad (9.60)$$

Substituting (9.28) into (9.55)

$$\dot{V}_{13} = k_2 \beta \tilde{\mathbf{q}}^T \mathbf{k}_p \dot{\tilde{\mathbf{q}}} = k_2 \beta \tilde{\mathbf{q}}^T \mathbf{k}_p \mathbf{G} \tilde{\boldsymbol{\omega}}$$
$$= k_2 \beta \tilde{\boldsymbol{\omega}}^T \left(\mathbf{G}^T \mathbf{k}_p \tilde{\mathbf{q}} \right) \quad (9.61)$$

Then, we have

$$\dot{V}_1 = \dot{V}_{11} + \dot{V}_{12} + \dot{V}_{13}$$
$$= k_2 \beta \tilde{\boldsymbol{\omega}}^T \left(\boldsymbol{\tau} + \mathbf{d}_\tau - EI\tilde{\mathbf{r}} \times \mathbf{d}_{xxx}(L, t) + T\tilde{\mathbf{r}} \times \mathbf{d}_x(L, t) + \mathbf{G}^T \mathbf{k}_p \tilde{\mathbf{q}} \right)$$
$$- k_2 \beta \int_\Omega y_t \left(EI y_{xxxx} - T y_{xx} \right) dx - k_2 \beta \int_\Omega z_t \left(EI z_{xxxx} - T z_{xx} \right) dx \quad (9.62)$$

Considering the term $\dot{V}_2$, we have

$$\dot{V}_2 = k_2 \beta EI \int_\Omega \left[y_{xx} y_{xxt} + z_{xx} z_{xxt} \right] dx + k_2 \beta T \int_\Omega \left[y_x y_{xt} + z_x z_{xt} \right] dx \quad (9.63)$$

Integrating (9.63) by parts with the boundary conditions, we obtain

$$\dot{V}_2 = k_2 \beta EI y_{xx} \, y_{xt}|_0^L - k_2 \beta EI y_{xxx} \, y_t|_0^L + k_2 \beta EI \int_\Omega y_{xxxx} y_t dx$$
$$+ k_2 \beta EI z_{xx} \, z_x|_0^L - k_1 \beta EI z_{xxx} \, z_t|_0^L + k_2 \beta EI \int_\Omega z_{xxxx} z_t dx$$
$$+ k_2 \beta T y_x \, y_t|_0^L - k_2 \beta T \int_\Omega y_{xx} y_t dx + k_1 \beta T z_x \, z_t|_0^L - k_2 \beta T \int_\Omega z_{xx} z_t dx$$
$$= k_2 \beta EI \int_\Omega y_{xxxx} y_t dx + k_2 \beta EI \int_\Omega z_{xxxx} z_t dx - k_2 \beta T \int_\Omega y_{xx} y_t dx$$
$$- k_2 \beta T \int_\Omega z_{xx} z_t dx - k_2 \beta EI y_{xxx}(L, t) y_t(L, t)$$
$$- k_2 \beta EI z_{xxx}(L, t) z_t(L, t) + k_2 \beta T y_x(L, t) y_t(L, t) + k_2 \beta T z_x(L, t) z_t(L, t)$$
(9.64)

Then we have

$$
\begin{aligned}
\dot{V}_1 + \dot{V}_2 &= k_2\beta\tilde{\boldsymbol{\omega}}^T \left(\boldsymbol{\tau} + \mathbf{d}_\tau - EI\tilde{\mathbf{r}} \times \mathbf{d}_{xxx}(L,t) + T\tilde{\mathbf{r}} \times \mathbf{d}_x(L,t)\right) \\
&\quad + k_2\beta\tilde{\boldsymbol{\omega}}^T \left(\mathbf{G}^T\mathbf{k}_p\tilde{\mathbf{q}}\right) - k_2\beta EI y_{xxx}(L,t)y_t(L,t) - k_2\beta EI z_{xxx}(L,t)z_t(L,t) \\
&\quad + k_2\beta T y_x(L,t)y_t(L,t) + k_2\beta T z_x(L,t)z_t(L,t) \\
&= k_2\beta\tilde{\boldsymbol{\omega}}^T \left(\boldsymbol{\tau} + \mathbf{d}_\tau + \mathbf{G}^T\mathbf{k}_p\tilde{\mathbf{q}}\right) - k_2\beta EI\mathbf{d}_t(L,t)^T\mathbf{d}_{xxx}(L,t) \\
&\quad - k_2\beta\tilde{\boldsymbol{\omega}}^T \left(EI\tilde{\mathbf{r}} \times \mathbf{d}_{xxx}(L,t) - T\tilde{\mathbf{r}} \times \mathbf{d}_x(L,t)\right) + k_2\beta T\mathbf{d}_t(L,t)^T\mathbf{d}_x(L,t)
\end{aligned}
\tag{9.65}
$$

By calculating, we get

$$
\begin{aligned}
&k_2 T\beta\tilde{\boldsymbol{\omega}}^T \left(\tilde{\mathbf{r}} \times \mathbf{d}_x(L,t)\right) - k_2 EI\beta\tilde{\boldsymbol{\omega}}^T \left(\tilde{\mathbf{r}} \times \mathbf{d}_{xxx}(L,t)\right) \\
&\quad - k_2\beta EI\mathbf{d}_t^T\mathbf{d}_{xxx}(L,t) + k_2\beta T\mathbf{d}_t^T\mathbf{d}_x(L,t) \\
&= -k_2 EI\beta\left[\tilde{\boldsymbol{\omega}} \times \tilde{\mathbf{r}}\right]^T\mathbf{d}_{xxx}(L,t) + k_2 T\beta\left[\tilde{\boldsymbol{\omega}} \times \tilde{\mathbf{r}}\right]^T\mathbf{d}_x(L,t) \\
&\quad - k_2\beta EI\mathbf{d}_t^T\mathbf{d}_{xxx}(L,t) + k_2\beta T\mathbf{d}_t^T\mathbf{d}_x(L,t) \\
&= -k_2\beta EI\left[\frac{d\tilde{\mathbf{r}}(L,t)}{dt}\right]^T\mathbf{d}_{xxx}(L,t) + k_2\beta T\left[\frac{d\tilde{\mathbf{r}}(L,t)}{dt}\right]^T\mathbf{d}_x(L,t)
\end{aligned}
\tag{9.66}
$$

Then (9.65) is rewritten as

$$
\begin{aligned}
\dot{V}_1 + \dot{V}_2 &= k_2\beta\tilde{\boldsymbol{\omega}}^T \left(\boldsymbol{\tau} + \mathbf{d}_\tau + \mathbf{G}^T\mathbf{k}_p\tilde{\mathbf{q}}\right) \\
&\quad - k_2\beta EI\left[\frac{d\tilde{\mathbf{r}}(L,t)}{dt}\right]^T\mathbf{d}_{xxx}(L,t) + k_2\beta T\left[\frac{d\tilde{\mathbf{r}}(L,t)}{dt}\right]^T\mathbf{d}_x(L,t)
\end{aligned}
\tag{9.67}
$$

Combining (9.36) and substituting (9.32) into (9.67) yields

$$
\begin{aligned}
\dot{V}_1 + \dot{V}_2 &= -k_2\beta\tilde{\boldsymbol{\omega}}^T\mathbf{k}_d\tilde{\boldsymbol{\omega}} - k_f k_2\beta\tilde{\boldsymbol{\omega}}^T\mathbf{G}^T\mathbf{G}\tilde{\boldsymbol{\omega}} \\
&\quad - \frac{1}{2}\beta EI k_1^2[\mathbf{d}_x(L,t)]^T\boldsymbol{\phi}^T\boldsymbol{\phi}\mathbf{d}_x(L,t) - k_2\beta\tilde{\boldsymbol{\omega}}^T\mathbf{k}_q\tilde{\mathbf{q}} \\
&\quad - \frac{1}{2}\beta EI\left[\frac{d\tilde{\mathbf{r}}(L,t)}{dt}\right]^T\boldsymbol{\phi}^T\boldsymbol{\phi}\frac{d\tilde{\mathbf{r}}(L,t)}{dt} + \frac{1}{2}\beta EI\mathbf{u}_0^T\mathbf{u}_0 \\
&\quad + \beta EI k_1 k_2[\mathbf{d}_x(L,t)]^T\boldsymbol{\phi}^T\boldsymbol{\phi}\mathbf{d}_{xxx}(L,t) + k_2\beta\tilde{\boldsymbol{\omega}}^T\tilde{\mathbf{d}}_\tau \\
&\quad + (k_2\beta T - \beta EI k_1)\left[\frac{d\tilde{\mathbf{r}}(L,t)}{dt}\right]^T\boldsymbol{\phi}^T\boldsymbol{\phi}\mathbf{d}_x(L,t) \\
&\quad - \frac{1}{2}\beta EI k_2^2[\mathbf{d}_{xxx}(L,t)]^T\boldsymbol{\phi}^T\boldsymbol{\phi}\mathbf{d}_{xxx}(L,t)
\end{aligned}
\tag{9.68}
$$

According to Lemma 2.4, we have

$$
\dot{V}_1 + \dot{V}_2 \leq -k_2\beta\tilde{\boldsymbol{\omega}}^T\mathbf{k}_d\tilde{\boldsymbol{\omega}} - k_f k_2\beta\tilde{\boldsymbol{\omega}}^T\mathbf{G}^T\mathbf{G}\tilde{\boldsymbol{\omega}}
$$
$$
+ \frac{1}{2}\beta EI\mathbf{u}_0^T\mathbf{u}_0 + \frac{1}{2\sigma_3}k_2\beta\tilde{\boldsymbol{\omega}}^T\tilde{\boldsymbol{\omega}} + \frac{1}{2}\sigma_3 k_2\beta\tilde{\mathbf{d}}_\tau^T\tilde{\mathbf{d}}_\tau
$$
$$
+ \frac{1}{2\sigma_2}\sqrt{k_2}\beta\tilde{\boldsymbol{\omega}}^T\mathbf{k}_q\tilde{\boldsymbol{\omega}} + \frac{1}{2}\sigma_2\sqrt{k_2}\beta\tilde{\mathbf{q}}^T\mathbf{k}_q\tilde{\mathbf{q}}
$$
$$
- \left(\frac{1}{2}\beta EI - \frac{1}{2\sigma_0}|k_2\beta T - k_1\beta EI|\right)\left[\frac{d\tilde{\mathbf{r}}(L,t)}{dt}\right]^T\boldsymbol{\phi}^T\boldsymbol{\phi}\frac{d\tilde{\mathbf{r}}(L,t)}{dt}
$$
$$
- \left(\frac{1}{2}\beta EI k_1^2 - \frac{1}{2}\sigma_0|k_2\beta T - k_1\beta EI|\right)[\mathbf{d}_x(L,t)]^T\boldsymbol{\phi}^T\boldsymbol{\phi}\mathbf{d}_x(L,t)
$$
$$
- \frac{1}{2}\beta EI k_2^2[\mathbf{d}_{xxx}(L,t)]^T\boldsymbol{\phi}^T\boldsymbol{\phi}\mathbf{d}_{xxx}(L,t) + \beta EI k_1 k_2[\mathbf{d}_x(L,t)]^T\boldsymbol{\phi}^T\boldsymbol{\phi}\mathbf{d}_{xxx}(L,t)
$$

$$(9.69)$$

Since

$$
\frac{d\mathbf{u}_0}{dt} = \boldsymbol{\phi}\left[2\tilde{\boldsymbol{\omega}} \times \mathbf{d}_t(L,t) + \dot{\tilde{\boldsymbol{\omega}}} \times \tilde{\mathbf{r}}(L,t) + \tilde{\boldsymbol{\omega}} \times (\tilde{\boldsymbol{\omega}} \times \tilde{\mathbf{r}}(L,t)) + \mathbf{d}_{tt}(L,t)\right]
$$
$$
+ \boldsymbol{\phi}\left[k_1\mathbf{d}_{xt}(L,t) + k_1\tilde{\boldsymbol{\omega}} \times \mathbf{d}_x(L,t) - k_2\mathbf{d}_{xxxt}(L,t) - k_2\tilde{\boldsymbol{\omega}} \times \mathbf{d}_{xxx}(L,t)\right]
$$
$$
= \boldsymbol{\phi}\left[\tilde{\boldsymbol{\varpi}}(L,t) + \mathbf{d}_{tt}(L,t) + k_1\mathbf{d}_{xt}(L,t) + k_1\tilde{\boldsymbol{\omega}} \times \mathbf{d}_x(L,t)\right]
$$
$$
- \boldsymbol{\phi}\left[k_2\mathbf{d}_{xxxt}(L,t) + k_2\tilde{\boldsymbol{\omega}} \times \mathbf{d}_{xxx}(L,t)\right]
$$

$$(9.70)$$

we obtain

$$
\dot{V}_3 = m\mathbf{u}_0^T\frac{d\mathbf{u}_0}{dt} = \mathbf{u}_0^T\left(\mathbf{u} + \mathbf{d}_u + EI\boldsymbol{\phi}\mathbf{d}_{xxx}(L,t) - T\boldsymbol{\phi}\mathbf{d}_x(L,t)\right)
$$
$$
+ m\mathbf{u}_0^T\boldsymbol{\phi}\left(k_1\mathbf{d}_{xt}(L,t) + k_1\tilde{\boldsymbol{\omega}} \times \mathbf{d}_x(L,t) - k_2\mathbf{d}_{xxxt}(L,t) - k_2\tilde{\boldsymbol{\omega}} \times \mathbf{d}_{xxx}(L,t)\right)
$$

$$(9.71)$$

Substituting (9.33) into (9.71), we have

$$
\dot{V}_3 = -\mathbf{u}_0^T\mathbf{k}_u\mathbf{u}_0 + \mathbf{u}_0^T\tilde{\mathbf{d}}_u
$$
$$
\leq -\mathbf{u}_0^T\mathbf{k}_u\mathbf{u}_0 + \frac{1}{2}\sigma_6\mathbf{u}_0^T\mathbf{u}_0 + \frac{1}{2\sigma_6}\tilde{\mathbf{d}}_u^T\tilde{\mathbf{d}}_u
$$

$$(9.72)$$

To go on, the term $\dot{V}_4$ can be written as

$$
\dot{V}_4 = B_1 + B_2 + B_3 + B_4 + B_5 + B_6 \tag{9.73}
$$

where

$$
B_1 = \alpha\rho\int_\Omega x\boldsymbol{\phi}^T\left[\tilde{\boldsymbol{\varpi}} + \mathbf{d}_{tt}\right]^T[\boldsymbol{\phi}\mathbf{d}_x]\,dx \tag{9.74}
$$

$$B_2 = \alpha\rho \int_\Omega x \left[\phi \frac{d\tilde{\mathbf{r}}}{dt} \right]^T \left[\phi \frac{d\mathbf{d}_x}{dt} \right] dx \tag{9.75}$$

$$B_3 = \alpha\rho \int_\Omega x \phi_1^T \left[\tilde{\varpi} + \mathbf{d}_{tt} \right]^T \left[\phi_2 \tilde{\mathbf{q}} \right] dx \tag{9.76}$$

$$B_4 = \alpha\rho \int_\Omega x \left[\phi_1 \frac{d\tilde{\mathbf{r}}}{dt} \right]^T \left[\phi_2 \dot{\tilde{\mathbf{q}}} \right] dx \tag{9.77}$$

$$B_5 = \alpha\tilde{\boldsymbol{\omega}}^T \mathbf{I_h} \dot{\tilde{\mathbf{q}}} \tag{9.78}$$

$$B_6 = \alpha\tilde{\mathbf{q}}^T \left(\mathbf{I_h}\dot{\tilde{\boldsymbol{\omega}}} + \tilde{\boldsymbol{\omega}} \times (\mathbf{I_h}\tilde{\boldsymbol{\omega}}) \right) \tag{9.79}$$

Using boundary conditions and integrating by parts, we obtain

$\dot{V}_4 = B_1 + B_2 + B_3 + B_4 + B_5 + B_6$

$\leq -\alpha EIL y_x(L,t) y_{xxx}(L,t) - \alpha EIL z_x(L,t) z_{xxx}(L,t) + \frac{1}{2}\alpha TL(z_x(L,t))^2 + \frac{\alpha}{2\sigma_5} EIL(y_{xxx}(L,t))^2$

$+ \frac{\alpha}{2\sigma_5} EIL(z_{xxx}(L,t))^2 + \frac{\alpha}{2\sigma_7} TL(y_x(L,t))^2 + \frac{1}{2}\alpha TL(y_x(L,t))^2$

$+ \frac{\alpha}{2\sigma_7} TL(z_x(L,t))^2 - \frac{3}{2}\alpha EI \int_\Omega (y_{xx})^2 dx - \frac{3}{2}\alpha EI \int_\Omega (z_{xx})^2 dx$

$- \frac{1}{2}\alpha T \int_\Omega (y_x)^2 dx - \frac{1}{2}\alpha T \int_\Omega (z_x)^2 dx - \left(\frac{1}{2}\alpha\rho - \sigma_4 \frac{\alpha\rho L}{2} \right) \int_\Omega \left[\phi \frac{d\tilde{\mathbf{r}}}{dt} \right]^T \left[\phi \frac{d\tilde{\mathbf{r}}}{dt} \right] dx$

$+ \frac{\alpha}{2}\sigma_8\rho L \int_\Omega \left[\phi \frac{d\tilde{\mathbf{r}}}{dt} \right]^T \left[\phi \frac{d\tilde{\mathbf{r}}}{dt} \right] dx + \frac{1}{2}\alpha L \left[\phi \frac{d\tilde{\mathbf{r}}(L,t)}{dt} \right]^T \left[\phi \frac{d\tilde{\mathbf{r}}(L,t)}{dt} \right] + \frac{1}{2}\sigma_9\alpha\tilde{\boldsymbol{\omega}}^T \mathbf{I_h}\tilde{\boldsymbol{\omega}}$

$+ \frac{\alpha}{2\sigma_{10}}\tilde{\boldsymbol{\omega}}^T \mathbf{k}_d\tilde{\boldsymbol{\omega}} + \frac{\alpha\rho}{2\sigma_4} L^2 \left(\tilde{\omega}_2^2 + \tilde{\omega}_3^2 \right) + \frac{\alpha}{2\sigma_{11}}\tilde{\mathbf{q}}^T\tilde{\mathbf{q}} - \alpha\tilde{\mathbf{q}}^T\mathbf{k}_q\tilde{\mathbf{q}} - \frac{1}{2}\alpha\tilde{\eta}\tilde{\mathbf{q}}^T\mathbf{k}_p\tilde{\mathbf{q}} + \frac{\alpha}{2}\sigma_{10}\tilde{\mathbf{q}}^T\mathbf{k}_d\tilde{\mathbf{q}}$

$+ \frac{\alpha}{4}k_f\tilde{\eta}^2\tilde{\mathbf{q}}^T\tilde{\mathbf{q}} + \frac{\alpha}{2}\sigma_5 EIL\tilde{q}_3^2 + \frac{\alpha}{2}\sigma_7 TL\tilde{q}_3^2 + \frac{\alpha}{2}\sigma_7 TL\tilde{q}_2^2 + \frac{\alpha}{2}\sigma_5 EIL\tilde{q}_2^2 + \frac{\alpha}{2}k_f[\mathbf{G}\tilde{\boldsymbol{\omega}}]^T\mathbf{G}\tilde{\boldsymbol{\omega}}$

$+ \frac{1}{2\sigma_9}\alpha[\mathbf{G}\tilde{\boldsymbol{\omega}}]^T\mathbf{I_h}[\mathbf{G}\tilde{\boldsymbol{\omega}}] + \frac{\alpha}{2}\sigma_{11}\tilde{\mathbf{d}}_\tau^T\tilde{\mathbf{d}}_\tau + \frac{\alpha}{2\sigma_8}\rho L \left([\mathbf{G}\tilde{\boldsymbol{\omega}}]^T\phi_2^T\phi_2[\mathbf{G}\tilde{\boldsymbol{\omega}}] \right) \tag{9.80}$

The last term of (9.51) is

$\dot{V}_o = \tilde{\mathbf{d}}_\tau^T\dot{\tilde{\mathbf{d}}}_\tau + \tilde{\mathbf{d}}_u^T\dot{\tilde{\mathbf{d}}}_u = -\tilde{\mathbf{d}}_\tau^T\dot{\hat{\mathbf{d}}}_\tau - \tilde{\mathbf{d}}_u^T\dot{\hat{\mathbf{d}}}_u + \tilde{\mathbf{d}}_\tau^T\dot{\mathbf{d}}_\tau + \tilde{\mathbf{d}}_u^T\dot{\mathbf{d}}_u$

$\leq -\tilde{\mathbf{d}}_\tau^T \left(\dot{\mathbf{z}}_1 + K_1 \left(\mathbf{I_h}\dot{\boldsymbol{\omega}} + \boldsymbol{\omega} \times (\mathbf{I_h}\boldsymbol{\omega}) \right) \right) - \tilde{\mathbf{d}}_u^T \left(\dot{\mathbf{z}}_2 + K_2 m\phi \left[\frac{d^2\mathbf{r}_L}{dt^2} \right] \right) + \sigma_{12}\tilde{\mathbf{d}}_\tau^T\tilde{\mathbf{d}}_\tau$

$+ \sigma_{13}\tilde{\mathbf{d}}_u^T\tilde{\mathbf{d}}_u + \frac{1}{\sigma_{12}}\dot{\mathbf{d}}_\tau^T\dot{\mathbf{d}}_\tau + \frac{1}{\sigma_{13}}\dot{\mathbf{d}}_u^T\dot{\mathbf{d}}_u$

$= -\tilde{\mathbf{d}}_\tau^T (K_1 - \sigma_{12}\mathbf{I}_3) \tilde{\mathbf{d}}_\tau - \tilde{\mathbf{d}}_u^T (K_2 - \sigma_{13}\mathbf{I}_3) \tilde{\mathbf{d}}_u + \frac{1}{\sigma_{12}}\dot{\mathbf{d}}_\tau^T\dot{\mathbf{d}}_\tau + \frac{1}{\sigma_{13}}\dot{\mathbf{d}}_u^T\dot{\mathbf{d}}_u \tag{9.81}$

Applying (9.69), (9.72), (9.80) and (9.81), we have

$$
\begin{aligned}
\dot{V} \leq & -\left(\frac{1}{2}\beta EI k_2^2 - \frac{\alpha}{2\sigma_5}EIL - \frac{1}{2\sigma_1}|k_1 k_2 \beta EI - \alpha EIL|\right)[\phi \mathbf{d}_{xxx}(L,t)]^T [\phi \mathbf{d}_{xxx}(L,t)] \\
& -\left(\frac{1}{2}\beta EI k_1^2 + \frac{\alpha EI}{L} - \frac{1}{2}\alpha TL - \frac{1}{2}\sigma_0|k_2\beta T - k_1\beta EI| - \frac{\alpha}{2\sigma_7}TL\right)[\phi \mathbf{d}_x(L,t)]^T [\phi \mathbf{d}_x(L,t)] \\
& -\left(-\frac{1}{2}\sigma_1|k_1 k_2 \beta EI - \alpha EIL|\right)[\phi \mathbf{d}_x(L,t)]^T [\phi \mathbf{d}_x(L,t)] - \frac{1}{2}\alpha EI\int_\Omega (y_{xx})^2 dx - \frac{1}{2}\alpha EI\int_\Omega (z_{xx})^2 dx \\
& -\frac{1}{2}\alpha T\int_\Omega (y_x)^2 dx - \frac{1}{2}\alpha\rho(1-\sigma_4 L - \sigma_8 L)\int_\Omega \left[\phi\frac{d\tilde{\mathbf{r}}}{dt}\right]^T \left[\phi\frac{d\tilde{\mathbf{r}}}{dt}\right] dx \\
& -\frac{1}{2}\left(\beta EI - \alpha L - \frac{1}{2\sigma_0}|k_2\beta T - k_1\beta EI|\right)\left[\frac{d\tilde{\mathbf{r}}(L,t)}{dt}\right]^T \phi^T \phi \frac{d\tilde{\mathbf{r}}(L,t)}{dt} - \frac{1}{2}\alpha T\int_\Omega (z_x)^2 dx \\
& -\tilde{\mathbf{q}}^T\left(\alpha\mathbf{k}_q + \frac{1}{2}\alpha\tilde{\eta}\mathbf{k}_p - \frac{\alpha}{2}\sigma_{10}\mathbf{k}_d\right)\tilde{\mathbf{q}} + \tilde{\mathbf{q}}^T\left(\frac{\alpha}{4}k_f\tilde{\eta}^2\mathbf{I}_3 + \frac{1}{2}\sigma_2\sqrt{k_2}\beta\mathbf{k}_q + \frac{\alpha}{2\sigma_{11}}\mathbf{I}_3\right)\tilde{\mathbf{q}} \\
& +\tilde{\mathbf{q}}^T\left(\frac{\alpha}{2}\sigma_5 EIL\phi^T\phi + \frac{\alpha}{2}\sigma_7 TL\phi^T\phi\right)\tilde{\mathbf{q}} - \frac{1}{2}\tilde{\omega}^T\left(2k_2\beta\mathbf{k}_d + \frac{1}{\sigma_3}k_2\beta\mathbf{I}_3 - \frac{\alpha\rho}{\sigma_4}L^2\phi^T\phi\right)\tilde{\omega} \\
& +\frac{1}{2}\tilde{\omega}^T\left(\frac{\alpha}{\sigma_{10}}\mathbf{k}_d + 2\sigma_9\alpha\mathbf{I_h} + \frac{1}{\sigma_2}\sqrt{k_2}\beta\mathbf{k}_q\right)\tilde{\omega} - [G\tilde{\omega}]^T\left(k_2 k_f\beta\mathbf{I}_3 - \frac{\alpha}{2\sigma_8}\rho L\phi_2^T\phi_2\right)[G\tilde{\omega}] \\
& -[G\tilde{\omega}]^T\left(-\frac{1}{2\sigma_9}\alpha\mathbf{I_h} - \frac{\alpha}{2}k_f\mathbf{I}_3\right)[G\tilde{\omega}] - \mathbf{u}_0^T\left(\mathbf{k}_u - \frac{1}{2}\beta EI\mathbf{I}_2 - \frac{1}{2}\sigma_6\mathbf{I}_2\right)\mathbf{u}_0 \\
& -\tilde{\mathbf{d}}_\tau^T\left(K_1 - \frac{\alpha}{2}\sigma_{11}\mathbf{I}_3 - \frac{1}{2}\sigma_3 k_2\beta\mathbf{I}_3 - \sigma_{12}\mathbf{I}_3\right)\tilde{\mathbf{d}}_\tau - \tilde{\mathbf{d}}_u^T\left(K_2 - \frac{1}{2\sigma_6}\mathbf{I}_2 - \sigma_{13}\mathbf{I}_2\right)\tilde{\mathbf{d}}_u + \varepsilon \quad (9.82)
\end{aligned}
$$

where $\varepsilon = \frac{1}{\sigma_{12}}(\bar{d}_{\tau x}^2 + \bar{d}_{\tau y}^2 + \bar{d}_{\tau z}^2) + \frac{1}{\sigma_{13}}(\bar{d}_{uy}^2 + \bar{d}_{uz}^2)$.

We design parameters k_1 k_2, $\mathbf{k}_p$, $\mathbf{k}_d$, $\mathbf{k}_q$, $\mathbf{k}_u$, K_1, K_2, α, β and σ_n $(n = 0...13)$ to satisfy the following conditions:

$$
\begin{aligned}
& \beta EI k_1^2 + \frac{2\alpha EI}{L} - \sigma_0|k_2\beta T - k_1\beta EI| - \alpha TL \\
& - \frac{\alpha}{\sigma_7}TL - \sigma_1|k_1 k_2 \beta EI - \alpha EIL| \geq 0
\end{aligned} \quad (9.83)
$$

$$
\beta EI k_2^2 - \frac{\alpha}{\sigma_5}EIL - \frac{1}{\sigma_1}|k_1 k_2 \beta EI - \alpha EIL| \geq 0 \quad (9.84)
$$

$$
\beta EI - \frac{1}{\sigma_0}|k_2\beta T - k_1\beta EI| - \alpha L \geq 0 \quad (9.85)
$$

$$
k_2 k_f\beta - \frac{1}{2\sigma_9}\alpha I_{h1} - \frac{\alpha}{2}k_f \geq 0 \quad (9.86)
$$

$$
k_2 k_f\beta - \frac{\alpha}{2\sigma_8}\rho L - \frac{1}{2\sigma_9}\alpha I_{hi} - \frac{\alpha}{2}k_f \geq 0 \quad (9.87)
$$

$$\delta_1 = 1 - \sigma_8 L - \sigma_4 L > 0 \tag{9.88}$$

$$\delta_2 = 2\alpha k_{q1} + \alpha\tilde{\eta}_0 k_{p1} - \alpha\sigma_{10}k_{d1} - \frac{1}{2}\alpha k_f \tilde{\eta}^2$$
$$- \sigma_2\sqrt{k_2}\beta k_{q1} - \frac{\alpha}{2\sigma_{11}} > 0 \tag{9.89}$$

$$\delta_{i+1} = 2\alpha k_{qi} + \alpha\tilde{\eta}_0 k_{pi} - \alpha\sigma_{10}k_{di} - \frac{1}{2}\alpha k_f \tilde{\eta}^2$$
$$- \sigma_2\sqrt{k_2}\beta k_{qi} - \alpha\sigma_7 T L - \alpha\sigma_5 EIL - \frac{\alpha}{2\sigma_{11}} > 0 \tag{9.90}$$

$$\delta_5 = 2k_2\beta k_{d1} + \frac{1}{\sigma_3}k_2\beta - \frac{\alpha}{\sigma_{10}}k_{d1}$$
$$- 2\sigma_9\alpha I_{h1} - \frac{1}{\sigma_2}\sqrt{k_2}\beta k_{q1} > 0 \tag{9.91}$$

$$\delta_{i+4} = 2k_2\beta k_{di} + \frac{1}{\sigma_3}k_2\beta - \frac{\alpha\rho}{\sigma_4}L^2 - \frac{\alpha}{\sigma_{10}}k_{di}$$
$$- 2\sigma_9\alpha I_{hi} - \frac{1}{\sigma_2}\sqrt{k_2}\beta k_{qi} > 0 \tag{9.92}$$

$$\delta_{j+7} = k_{uj} - \frac{1}{2}\beta T - \frac{1}{2}\sigma_6 > 0 \tag{9.93}$$

$$\delta_{j+9} = K_{2j} - \frac{1}{2\sigma_6} - \sigma_{13} > 0 \tag{9.94}$$

$$\delta_{m+11} = K_{1m} - \frac{\alpha}{2}\sigma_{11} - \frac{1}{2}\sigma_3 k_2\beta - \sigma_{12} > 0 \tag{9.95}$$

where $m = 1, 2, 3$, $j = 1, 2$, $i = 2, 3$.

We can obtain

$$\dot{V} \le -\lambda_1 (V_1 + V_2 + V_3 + V_o) + \varepsilon \tag{9.96}$$

where $\lambda_1 = \min\left(\frac{\alpha\delta_1}{\beta k_1}, \frac{2\delta_{i+1}}{\beta k_1 k_{pi}}, \frac{\delta_{i+4}}{\beta k_1 I_{hi}}, \frac{\alpha}{\beta k_1}, \frac{2\delta_{j+7}}{m}, 2\delta_{j+9}, 2\delta_{m+11}\right)$

Combining (9.50) and (9.96), we have

$$\dot{V} \le -\lambda V + \varepsilon \tag{9.97}$$

where $\lambda = \lambda_1/\alpha_3 > 0$.

Then, multiply Eq. (9.97), by $e^{\lambda t}$, we obtain

$$\frac{d}{dt}\left((V(t)e^{\lambda t})\right) \le \varepsilon e^{\lambda t} \tag{9.98}$$

Integration of the above inequalities, we obtain

$$V(t) \le V(0)e^{-\lambda t} + \frac{\varepsilon}{\lambda}\left(1 - e^{-\lambda t}\right) \le V(0)e^{-\lambda t} + \frac{\varepsilon}{\lambda} \tag{9.99}$$

We can conclude that $V(t)$ is bounded from (9.99).
According to Lemma 2.5, we have

$$\frac{\beta k_2 T}{2L} y^2(x, t) \leq \frac{\beta k_2 T}{2} \int_0^L (y_x)^2 dx \leq V_2(t) \leq \frac{V(t)}{\alpha_2} \tag{9.100}$$

$$\frac{\beta k_2 T}{2L} z^2(x, t) \leq \frac{\beta k_2 T}{2} \int_0^L (z_x)^2 dx \leq V_2(t) \leq \frac{V(t)}{\alpha_2} \tag{9.101}$$

Therefore we obtain $y(x, t)$ and $z(x, t)$ are uniformly bounded as follows

$$|y(x, t)| \leq \sqrt{\frac{2L}{\beta k_2 T \alpha_2} V(t)} \leq \sqrt{\frac{2L}{\beta k_2 T \alpha_2} \left(V(0)e^{-\lambda t} + \frac{\varepsilon}{\lambda} \right)} \tag{9.102}$$

$$|z(x, t)| \leq \sqrt{\frac{2L}{\beta k_2 T \alpha_2} V(t)} \leq \sqrt{\frac{2L}{\beta k_2 T \alpha_2} \left(V(0)e^{-\lambda t} + \frac{\varepsilon}{\lambda} \right)} \tag{9.103}$$

$\forall (x, t) \in [0, L] \times [0, \infty)$.
Similarly, we have

$$|\tilde{q}_i| \leq \sqrt{\frac{2V(t)}{\beta k_2 \alpha_2 \lambda_{\min}(\mathbf{k}_p)}} \leq \sqrt{\frac{2}{\beta k_2 \alpha_2 \lambda_{\min}(\mathbf{k}_p)} \left(V(0)e^{-\lambda t} + \frac{\varepsilon}{\lambda} \right)} \quad (i = 1, 2, 3) \tag{9.104}$$

It follows that,

$$\lim_{t \to \infty} |y(x, t)| \leq \sqrt{\frac{2L\varepsilon}{\beta k_2 T \alpha_2 \lambda}} \tag{9.105}$$

$$\lim_{t \to \infty} |z(x, t)| \leq \sqrt{\frac{2L\varepsilon}{\beta k_2 T \alpha_2 \lambda}} \tag{9.106}$$

$\forall (x, t) \in [0, L] \times [0, \infty)$£¬ and

$$\lim_{t \to \infty} |\tilde{q}_i| \leq \sqrt{\frac{2\varepsilon}{\beta k_2 \alpha_2 \lambda_{\min}(\mathbf{k}_p) \lambda}} \quad (i = 1, 2, 3) \tag{9.107}$$

$\forall t \in [0, \infty)$. Therefore, $y(x, t)$, $z(x, t)$ and $\tilde{q}_i$ $(i = 1, 2, 3)$ are uniformly ultimate bounded.

Appendix 2: Simulation Program

```
1    % Model based boundary control
2    close all;
3    clear all;
4    nx=20;
5    nt=100*10^3;
6
7    L=1;
8    tmax=15;
9    Ttr=80;                %80
10   h-L/(nx-1);            %Δ L
11   ts=tmax/(nt-1);        %Sampling time
12
13   % Parameters
14   M=1;EI=8;
15   Ih1=0.15;Ih2=0.2;Ih3=0.2;
16   T=5;  %Tension
17   rho=0.2;beta=0.5;
18   k2=15;k3=15;
19   kp1=15;kd1=15;
20   kp2=25;kd2=15;
21   kp3=25;kd3=15;
22   kf1=15;kq1=15;
23
24   r=0.2;
25
26   Ttr1=200;      %200
27   fx=zeros(nx,nt);
28   fy=zeros(nx,nt);
29   fz=zeros(nx,nt);
30
31   udy=zeros(nt,1);
32   udz=zeros(nt,1);
33
34   for j=1:nt
35   for i=1:nx
36   fx(i,j)=(1+2*sin(0.2*(i-1)*h*j*ts)+3*sin(0.3*(i-1)*h*j*ts)
37          +5*sin(0.5*(i-1)*h*j*ts))*(i-1)*h/L/5;
38   fy(i,j)=(1+2*sin(0.2*(i-1)*h*j*ts)+3*sin(0.3*(i-1)*h*j*ts)
39          +5*sin(0.5*(i-1)*h*j*ts))*(i-1)*h/L/5;
40   fz(i,j)=(2+2*sin(0.2*(i-1)*h*j*ts)+3*sin(0.3*(i-1)*h*j*ts)
41          +5*sin(0.5*(i-1)*h*j*ts))*(i-1)*h/L/5;
42   fx(i,j)=0;
43   fy(i,j)=0;
44   fz(i,j)=0;
45   end
46
47   udy(j)=(1+0.5*sin(0.5*j*ts)+1*sin(1*j*ts))/5;
48   udz(j)=(1+0.5*sin(0.5*j*ts)+1*sin(1*j*ts))/5;
49
50   tdx(j)=(1+0.5*sin(0.5*j*ts)+1*sin(0.1*j*ts))/5;
51   tdy(j)=(1+1*sin(0.1*j*ts)+3*sin(0.5*j*ts)+5*sin(1*j*ts))/5;
52   tdz(j)=(1+1*sin(0.1*j*ts)+3*sin(0.5*j*ts)+5*sin(1*j*ts))/5;
53
54   end
55   ud_bar=0;
56   td_bar=0;
54
55   %Create matrixes to save state data
```

```
56    x=zeros(nx,nt);
57    y=zeros(nx,nt);
58    z=zeros(nx,nt);
59    %Create matrixes to save control inputs
60    ux=zeros(nt,1);
61    uy=zeros(nt,1);
62    uz=zeros(nt,1);
63
64    x_3D=zeros(Ttr,nx);
65    y_3D=zeros(Ttr,nx);
66    z_3D=zeros(Ttr,nx);
67
68    xl_2D=zeros(nt,1);
69    yl_2D=zeros(nt,1);
70    zl_2D=zeros(nt,1);
71
72
73    xll_2D=zeros(nt,1);%L/2
74    yll_2D=zeros(nt,1);%L/2
75    zll_2D=zeros(nt,1);%L/2
76
77
78
79    ux_2D=zeros(Ttr1,1);
80    uy_2D=zeros(Ttr1,1);
81    uz_2D=zeros(Ttr1,1);
82    %%%%%%
83    wx=zeros(nt,1);
84    wy=zeros(nt,1);
85    wz=zeros(nt,1);
86    qx=zeros(nt,1);
87    qy=zeros(nt,1);
88    qz=zeros(nt,1);
89
90
91    dty=zeros(nx,nt);
92    dtz=zeros(nx,nt);
93
94
95    for j=1:nt
96    wbar_x(j)=0;
97    wbar_y(j)=0;
98    wbar_z(j)=0;
99    dtwx(j)=0;
100   dtwy(j)=0;
101   dtwz(j)=0;
102
103   theta1=0;theta2=pi/3;theta3=pi/3;
104   eta(j)=1;
105
106   etad(j)=cos(theta1/2)*cos(theta2/2)*cos(theta3/2)
107           +sin(theta1/2)*sin(theta2/2)*sin(theta3/2);
108   qxd(j)=sin(theta1/2)*cos(theta2/2)*cos(theta3/2)
109          -cos(theta1/2)*sin(theta2/2)*sin(theta3/2);
110   qyd(j)=cos(theta1/2)*sin(theta2/2)*cos(theta3/2)
111          +sin(theta1/2)*cos(theta2/2)*sin(theta3/2);
112   qzd(j)=cos(theta1/2)*cos(theta2/2)*sin(theta3/2)
113          -sin(theta1/2)*sin(theta2/2)*cos(theta3/2);
114   wxd(j)=0;
115   wyd(j)=0;
116   wzd(j)=0;
117
118   dex(j)=0;
```

```
119   dey(j)=0;
120   dez(j)=0;
121   taux(j)=0;% x m
122   tauy(j)=0;% x m
123   tauz(j)=0;% z
120
121
122   %%%%%%%%%%%
123   tdp1(j)=0;
124   tdp2(j)=0;
125   tdp3(j)=0;
126   Tdp=0;
127   udp1(j)=0;
128   udp2(j)=0;
129   z11(j)=0;
130   z12(j)=0;
131   z13(j)=0;
132   z21(j)=0;
133   z22(j)=0;
134   end
135
136   for i=1:nx
137   y(i,1)=0.25*(i-1)*h/L;
138   z(i,1)=0.25*(i-1)*h/L;
139   end
140
141   y(:,2)=y(:,1);
142   z(:,2)=z(:,1);
143   %
144   ysss1=0;
145   zsss1=0;
146   ys1=0;
147   zs1=0;
148
149   %  Main cycle   %
150   for j=2:nt-1
151
152
153   yss0=(y(3,j)-2*y(2,j)+y(1,j))/h^2;
154   zss0=(z(3,j)-2*z(2,j)+z(1,j))/h^2;
155   R1=-(Ih2-Ih3)*wy(j)*wz(j);
156   R2=(Ih1-Ih3)*wx(j)*wz(j)+EI*zss0+T*z(nx,j);
157   R3=-(Ih1-Ih2)*wx(j)*wy(j)+EI*(-yss0)+T*(-y(nx,j));
158
159   wx(j+1)=wx(j)+ts/Ih1*(taux(j)+tdx(j)-R1);
160   wy(j+1)=wy(j)+ts/Ih2*(tauy(j)+tdy(j)-R2);
161   wz(j+1)=wz(j)+ts/Ih3*(tauz(j)+tdz(j)-R3);
162
163   dtwx(j+1)=(wx(j+1)-wx(j))/ts;
164   dtwy(j+1)=(wy(j+1)-wy(j))/ts;
165   dtwz(j+1)=(wz(j+1)-wz(j))/ts;
166
167   Gx=0.5*(eta(j)*wx(j)-qz(j)*wy(j)+qy(j)*wz(j));
168   Gy=0.5*(eta(j)*wy(j)+qz(j)*wx(j)-qx(j)*wz(j));
169   Gz=0.5*(eta(j)*wz(j)-qy(j)*wx(j)+qx(j)*wy(j));
170
171   eta(j+1)=eta(j)-0.5*(qx(j)*wx(j)+qy(j)*wy(j)+qz(j)*wz(j))*ts;
172   qx(j+1)=qx(j)+Gx*ts;
173   qy(j+1)=qy(j)+Gy*ts;
174   qz(j+1)=qz(j)+Gz*ts;
175
176   eeta(j+1)=eta(j+1)*etad(j+1)+qxd(j+1)*qx(j+1)++qyd(j+1)*qy
177          (j+1)+qzd(j+1)*qz(j+1);
```

```
178    ex(j+1)=etad(j+1)*qx(j+1)-eta(j+1)*qxd(j+1)-qz(j+1)*qyd(j+1)
179        +qy(j+1)*qzd(j+1);
180    ey(j+1)=etad(j+1)*qy(j+1)-eta(j+1)*qyd(j+1)+qz(j+1)*qxd(j+1)
181        -qx(j+1)*qzd(j+1);
182    ez(j+1)=etad(j+1)*qz(j+1)-eta(j+1)*qzd(j+1)-qy(j+1)*qxd(j+1)
183        +qx(j+1)*qyd(j+1);
184    dex(j+1)=wx(j+1)-wxd(j+1);
185    dey(j+1)=wy(j+1)-wyd(j+1);
186    dez(j+1)=wz(j+1)-wzd(j+1);
187
188    W=[wx(j+1),wy(j+1),wz(j+1)]';
189    SW=[0,-wz(j+1),wy(j+1);wz(j+1),0,-wx(j+1);-wy(j+1),wx(j+1),0];
190    Q=[ex(j+1),ey(j+1),ez(j+1)]';
191    SQ=[0,-ez(j+1),ey(j+1);ez(j+1),0,-ex(j+1);-ey(j+1),ex(j+1),0];
192    G=0.5*(SQ+eeta(j+1)*eye(3,3));
189
190    kf=15;
191    kp1=15;kp2=10;kp3=15;
192    kd1=15;kd2=10;kd3=15;
193    kq1=15;kq2=10;kq3=15;
194    Kp=[kp1,0,0;0,kp2,0;0,0,kp3];
195    Kd=[kd1,0,0;0,kd2,0;0,0,kd3];
196    Kq=[kq1,0,0;0,kq2,0;0,0,kq3];
197
198
199
200    Tau=-G'*Kp*Q-Kd*W-kf*G'*G*W-Kq*Q-Tdp;
201
202    taux(j+1)=Tau(1);% x  m
203    tauy(j+1)=Tau(2);% x  m
204    tauz(j+1)=Tau(3);% z
205
206
207
208    for  i=3:nx-2
209
210    ys=(y(i,j)-y(i-1,j))/h;
211    zs=(z(i,j)-z(i-1,j))/h;
212
213
214    yss=(y(i+1,j)-2*y(i,j)+y(i-1,j))/h^2;
215    zss=(z(i+1,j)-2*z(i,j)+z(i-1,j))/h^2;
216
217    yssss=(y(i+2,j)-4*y(i+1,j)+6*y(i,j)-4*y(i-1,j)+y(i-2,j))/h^4;
218    zssss=(z(i+2,j)-4*z(i+1,j)+6*z(i,j)-4*z(i-1,j)+z(i-2,j))/h^4;
219
220
221    wbar_y(j)=2*(-wx(j)*dtz(i,j))-(wx(j)^2+wz(j)^2)*y(i,j)
222        +(wy(j)*wz(j)-dtwx(j))*z(i,j) \ldots
223    +h*(i-1)*(wx(j)*wy(j)+dtwz(j));
224    wbar_z(j)=2*(wx(j)*dty(i,j))-(wy(j)*wz(j)+dtwx(j))*y(i,j)
225        -(wx(j)^2+wy(j)^2)*z(i,j) \ldots
226    +h*(i-1)*(wx(j)*wz(j)-dtwy(j));
227
228    S1=T*yss-EI*yssss-rho*wbar_y(j)+fy(i,j);
229    S2=T*zss-EI*zssss-rho*wbar_z(j)+fz(i,j);
230
231    y(i,j+1)=2*y(i,j)-y(i,j-1)+ts^2*S1/rho;
232    z(i,j+1)=2*z(i,j)-z(i,j-1)+ts^2*S2/rho;
233
234    dty(i,j+1)=(y(i,j+1)-y(i,j))/ts;
235    dtz(i,j+1)=(z(i,j+1)-z(i,j))/ts;
236
```

```
237
238
239     end
238
239     y(2,j+1)=(y(1,j+1)+y(3,j+1))/2;
240     z(2,j+1)=(z(1,j+1)+z(3,j+1))/2;
241
242     wbar_yl(j)=2*(-wx(j)*dtz(nx,j))-(wx(j)^2+wz(j)^2)*y(nx,j)
243                +(wy(j)*wz(j)-dtwx(j))*z(nx,j)  \ldots
244     +L*(wx(j)*wy(j)+dtwz(j));
245     wbar_zl(j)=2*(wx(j)*dty(nx,j))-(wy(j)*wz(j)+dtwx(j))*y(nx,j)
246                -(wx(j)^2+wy(j)^2)*z(nx,j)  \ldots
247     +L*(wx(j)*wz(j)-dtwy(j));
248
249     SL1=T*ysl-EI*ysssl+M*wbar_yl(j);
250     SL2=T*zsl-EI*zsssl+M*wbar_zl(j);
251
252     y(nx,j+1)=2*y(nx,j)-y(nx,j-1)+(-SL1+udy(j)+uy(j))*ts^2/M;
253     z(nx,j+1)=2*z(nx,j)-z(nx,j-1)+(-SL2+udz(j)+uz(j))*ts^2/M;
254
255     y(nx-1,j+1)=(y(nx,j+1)+y(nx-2,j+1))/2;
256     z(nx-1,j+1)=(z(nx,j+1)+z(nx-2,j+1))/2;
257
258     dty(nx,j+1)=(y(nx,j+1)-y(nx,j))/ts;
259     dtz(nx,j+1)=(z(nx,j+1)-z(nx,j))/ts;
260     dty(nx-1,j+1)=(y(nx-1,j+1)-y(nx-1,j))/ts;
261     dtz(nx-1,j+1)=(z(nx-1,j+1)-z(nx-1,j))/ts;
262
263
264
265     dysl=(y(nx,j+1)-y(nx-1,j+1)-y(nx,j)+y(nx-1,j))/(ts*h);
266     dzsl=(z(nx,j+1)-z(nx-1,j+1)-z(nx,j)+z(nx-1,j))/(ts*h);
267
268     dyl=(y(nx,j+1)-y(nx,j))/ts;
269     dzl=(z(nx,j+1)-z(nx,j))/ts;
270
271     ysl=(y(nx,j+1)-y(nx-1,j+1))/h;
272     zsl=(z(nx,j+1)-z(nx-1,j+1))/h;
273
274
275     ysssl=(-y(nx,j+1)+2*y(nx-1,j+1)-y(nx-2,j+1))/h^3;
276     zsssl=(-z(nx,j+1)+2*z(nx-1,j+1)-z(nx-2,j+1))/h^3;
277     dysssl=((-y(nx,j+1)+2*y(nx-1,j+1)-y(nx-2,j+1))-(-y(nx,j)
278            +2*y(nx-1,j)-y(nx-2,j)))/(ts*h^3);
279     dzsssl=((-z(nx,j+1)+2*z(nx-1,j+1)-z(nx-2,j+1))-(-z(nx,j)
280            +2*z(nx-1,j)-z(nx-2,j)))/(ts*h^3);
281
282     yl=y(nx,j+1);
283     zl=z(nx,j+1);
280
281
282
283
284     Rl=[0,yl,zl]';
285     dDl=[0,dyl,dzl]';
286     Dxl=[0,ysl,zsl]';
287     Dxxxl=[0,ysssl,zsssl]';
288     wdxl=SW*Dxl;
289     wdxxxl=SW*Dxxxl;
290     fai=[0,1,0;0,0,1];
291
292     k1=2;
293     k2=2;
```

```
294    ku1=40;
295    ku2=40;
296
297    u0=fai*(dD1+SW*R1+k1*Dx1-k2*Dxxx1);
298
299    uy(j+1)=-ku1*u0(1)-EI*ysss1+T*ysl-M*(k1*dysl+k1*wdx1(2)
300           -k2*dysss1-k2*wdxxx1(2))-1*udp1(j+1);
301    uz(j+1)=-ku2*u0(2)-EI*zsss1+T*zsl-M*(k1*dzsl+k1*wdx1(3)
302           -k2*dzsss1-k2*wdxxx1(3))-1*udp2(j+1);
303
304
305
306    %  %%%%%%%%%%%%%%%%%%%%
307
308    K11=40;K12=40;K13=40;
309    K21=20;K22=10;
310    Mx=EI*[0,zss0,-yss0]';
311    yl1=y(nx,j);
312    zl1=z(nx,j);
313    Nx=T*[0,zl1,-yl1]';
314
315    z11(j+1)=z11(j)+ts*(-K11*(taux(j)-Mx(1)-Nx(1))-K11*tdp1(j));
316    tdp1(j+1)=z11(j+1)+K11*Ih1*wx(j+1);
317
318    z12(j+1)=z12(j)+ts*(-K12*(tauy(j)-Mx(2)-Nx(2))-K12*tdp2(j));
319    tdp2(j+1)=z12(j+1)+K12*Ih2*wy(j+1);
320
321    z13(j+1)=z13(j)+ts*(-K13*(tauz(j)-Mx(3)-Nx(3))-K13*tdp3(j));
322    tdp3(j+1)=z13(j+1)+K13*Ih3*wz(j+1);
323
324    ysl1=(y(nx,j)-y(nx-1,j))/h;
325    zsl1=(z(nx,j)-z(nx-1,j))/h;
326    ysss11=(-y(nx,j)+2*y(nx-1,j)-y(nx-2,j))/h^3;
327    zsss11=(-z(nx,j)+2*z(nx-1,j)-z(nx-2,j))/h^3;
328
329
330    Dxl1=[0,ysl1,zsl1]';
331    Dxxxl1=[0,ysss11,zsss11]';
332    MMx=EI*fai*Dxxxl1-T*fai*Dxl1;
333    RR1=[L,yl,zl]';
334    DRR1=fai*(dD1+SW*RR1);
335
336    z21(j+1)=z21(j)+ts*(-K21*(uy(j)+MMx(1))-K21*udp1(j));
337    udp1(j+1)=z21(j+1)+K21*M*DRR1(1);
338
339    z22(j+1)=z22(j)+ts*(-K22*(uz(j)+MMx(2))-K22*udp2(j));
340    udp2(j+1)=z22(j+1)+K22*M*DRR1(2);
341
342    Tdp=[tdp1(j+1),tdp2(j+1),tdp3(j+1)]';
343
344    %%%%%%%%%%%%%%%%%%%%%%%%
343
344
345
346
347    x1_2D(j+1)=x(nx,j+1);
348    y1_2D(j+1)=y(nx,j+1);
349    z1_2D(j+1)=z(nx,j+1);
350
351
352    xll_2D(j+1)=x(nx/2,j+1);%L/2
353    yll_2D(j+1)=y(nx/2,j+1);%L/2
354    zll_2D(j+1)=z(nx/2,j+1);%L/2
```

```
355
356    if mod(j,nt/Ttr)==0     %
357    x_3D(1+j*Ttr/nt,:)=x(:,j)';
358    y_3D(1+j*Ttr/nt,:)=y(:,j)';
359    z_3D(1+j*Ttr/nt,:)=z(:,j)';
360    end
361    if mod(j,nt/Ttr1)==0    %
362    ux_2D(1+j*Ttr1/nt)=ux(j+1);
363    uy_2D(1+j*Ttr1/nt)=uy(j+1);
364    uz_2D(1+j*Ttr1/nt)=uz(j+1);
365    end
366    end
367
368
369
370    x_3D(1,:)=x(:,1)';
371    y_3D(1,:)=y(:,1)';
372    z_3D(1,:)=z(:,1)';
373
374    x1_2D(1)=x(nx,1);
375    x1_2D(2)=x1_2D(1);
376    y1_2D(1)=y(nx,1);
377    y1_2D(2)=y1_2D(1);
378    z1_2D(1)=z(nx,1);
379    z1_2D(2)=z1_2D(1);
380
381
382
383    x11_2D(1)=x(nx/2,1);%L/2
384    x11_2D(2)=x11_2D(1);%L/2
385    y11_2D(1)=y(nx/2,1);%L/2
386    y11_2D(2)=y11_2D(1);%L/2
387    z11_2D(1)=z(nx/2,1);%L/2
388    z11_2D(2)=z11_2D(1);%L/2
389
390
391    thx_2=0.52;thx_1=0.52;%z45  x30
392
393    thz_2=0.79;thz_1=0.79;
394
395    x1_1=0;x1_2=0;
396    y1_1=0;y1_2=0;
397
398    xa_1=0.61;xa_2=0.61;
399    ya_1=0.35;ya_2=0.35;
400    za_1=0.71;za_2=0.71;
401
402    x0=linspace(0,L,nx);
403    t_tr=linspace(0,tmax,Ttr);
404
405
406
407    t=linspace(0,tmax,nt);
408    figure(1)
409    plot(t,ex,'b','linewidth',2);
410    hold on
411    plot(t,ey,'r','linewidth',2);
412    hold on
413    plot(t,ez,'g','linewidth',2);
414    hold on
415    plot(t,eeta,'k','linewidth',2);
416    xlabel('time [s]','fontname','latex','fontsize',10);
417    h=legend('$\tilde{q}_1$','$\tilde{q}_2$','$\tilde{q}_3$',
```

```
418                    '$\tilde{\eta}$');
419     set(h,'Interpreter','latex','fontsize',10)
420     title('Orientation tracking error of the flexible manipulator')
421
422
423     figure(2);
424     plot(t,wx,'b','linewidth',2);
425     hold on
426     plot(t,wy,'r','linewidth',2);
427     hold on
428     plot(t,wz,'g','linewidth',2);
429     xlabel('time [s]','fontname','latex','fontsize',12);
430     ylabel('[rad/s]','fontname','latex','fontsize',12);
431     h=legend('$\tilde{\omega}_1$','$\tilde{\omega}_2$',
432                 '$\tilde{\omega}_3$');
433     set(h,'Interpreter','latex','fontsize',12')
434     title('Angular velocity tracking error of the flexible ...
            manipulatorr')
435
436
437
438     figure(3);
439     subplot(311)
440     plot(linspace(0,tmax,nt),taux,'b','linewidth',2);
441     axis([0 15 -4 0.5]);
442     xlabel('time [s]','fontname','latex','fontsize',12);
443     ylabel('\tau_{x} [Nm]','fontname','latex','fontsize',12);
444     title('Control input ...
            \tau_{x}','fontname','latex','fontsize',12);
445
446     subplot(312)
447     plot(linspace(0,tmax,nt),tauy,'b','linewidth',2);
448     xlabel('time [s]','fontname','latex','fontsize',12);
449     ylabel('\tau_{y} [Nm]','fontname','latex','fontsize',12);
450     title('Control input ...
            \tau_{y}','fontname','latex','fontsize',12);
451     subplot(313)
452     plot(linspace(0,tmax,nt),tauz,'b','linewidth',2);
453     xlabel('time [s]','fontname','latex','fontsize',12);
454     ylabel('\tau_{z} [Nm]','fontname','latex','fontsize',12);
455     title('Control input ...
            \tau_{y}','fontname','latex','fontsize',12);
456
457     figure(4);
458     surf(x0,t_tr,y_3D);view(45,30);
459     title({'Displacement of the beam in Y direction'});
460     ylabel('time [s]');xlabel('x [m]');zlabel('y(x,t)[m]');
461
462
463     figure(5);
464     surf(x0,t_tr,z_3D);view(45,30);
465     title({'Displacement of the beam in Z direction'});
466     ylabel('time [s]');xlabel('x [m]');zlabel('z(x,t)[m]');
467
468
469     figure(6);
470     subplot(211);
471     plot(linspace(0,tmax,nt),yl_2D,'b','linewidth',2);
472     xlabel('time [s]');ylabel('y(L,t) [m]');
473     title('End-point deflection y(L,t)');
474     subplot(212);
475     plot(linspace(0,tmax,nt),zl_2D,'b','linewidth',2);
476     xlabel('time [s]');ylabel('z(L,t) [m]');
```

```
477   title('End-point deflection z(L,t)');
478
479   figure(7);
480   subplot(211)
481   plot(linspace(0,tmax,Ttr1),uy_2D,'b','linewidth',2);
482   xlabel('time [s]','fontname','latex','fontsize',12);
483   ylabel('u_{y} [N]','fontname','latex','fontsize',12);
484   title('Control input u_y','fontname','latex','fontsize',12);
485   subplot(212)
486   plot(linspace(0,tmax,Ttr1),uz_2D,'b','linewidth',2);
487   xlabel('time [s]','fontname','latex','fontsize',12);
488   ylabel('u_{z} [N]','fontname','latex','fontsize',12);
489   title('Control input u_z','fontname','latex','fontsize',12);
488
409

490   figure(8);
491   subplot(311)
492   plot(linspace(0,tmax,nt),tdp1,'r:',linspace(0,tmax,nt),
493        tdx,'b','linewidth',2);
494   xlabel('time [s]');
495   ylabel('[Nm] ');
496   title('d_{\tau x} and its estimation');
497   h=legend('$\hat{d}_{\tau x}$','$d_{\tau x}$');
498   set(h,'Interpreter','latex','fontsize',14')
499   subplot(312)
500   plot(linspace(0,tmax,nt),tdp2,'r:',linspace(0,tmax,nt),
501        tdy,'b','linewidth',2);
502   xlabel('time [s]');
503   ylabel('[Nm] ');
504   title('d_{\tau y} and its estimation');
505   h=legend('$\hat{d}_{\tau y}$','$d_{\tau y}$');
506   set(h,'Interpreter','latex','fontsize',14')
507   subplot(313)
508   plot(linspace(0,tmax,nt),tdp3,'r:',linspace(0,tmax,nt),
509        tdz,'b','linewidth',2);
510   xlabel('time [s]');
511   ylabel('[Nm] ');
512   title('d_{{\tau}z} and its estimation');
513   h=legend('$\hat{d}_{{\tau}z}$','$d_{{\tau}z}$');
514   set(h,'Interpreter','latex','fontsize',14')
515
516

517   figure(9);
518   subplot(211)
519   plot(linspace(0,tmax,nt),udp1,'r:',linspace(0,tmax,nt),
520        udy,'b','linewidth',2);
521   xlabel('time [s]');
522   ylabel('[Nm] ');
523   title('d_{uy} and its estimation');
524   h=legend('$\hat{d}_{uy}$','$d_{uy}$');
525   set(h,'Interpreter','latex','fontsize',14')
526   subplot(212)
527   plot(linspace(0,tmax,nt),udp2,'r:',linspace(0,tmax,nt),
528        udz,'b','linewidth',2);
529   axis([0 15 -0.5 0.6]);
530   xlabel('time [s]');
531   ylabel('[Nm] ');
532   title('d_{uz} and its estimation');
533   h=legend('$\hat{d}_{u}$','$d_{uz}$');
534   set(h,'Interpreter','latex','fontsize',14')
```

References

1. Do KD, Pan J (2009) Boundary control of three-dimensional inextensible marine risers. J Sound Vib 327(3):299–321
2. Ge SS, Lee TH, Zhu G (1998) Improving regulation of a single-link flexible manipulator with strain feedback. IEEE Trans Robot Autom 14(1):179–185
3. He W, Ge SS (2015) Vibration control of a flexible beam with output constraint. IEEE Trans Ind Electron 62(8):5023–5030
4. He W, Yang C, Zhu J, Liu J-K, He X (2017) Active vibration control of a nonlinear three-dimensional euler-bernoulli beam. J Vib Control 23(19):3196–3215
5. Liu Z, Liu J, He W (2017) Partial differential equation boundary control of a flexible manipulator with input saturation. Int J Syst Sci 48(1):53–62
6. Nguyen TL, Do KD, Pan J (2013) Boundary control of coupled nonlinear three dimensional marine risers. J Mar Sci Appl 12(1):72–88
7. Wang H, Zhang R, Chen W, Liang X, Pfeifer R (2016) Shape detection algorithm for soft manipulator based on fiber bragg gratings. IEEE/ASME Trans Mechatron 21(6):2977–2982
8. Zhang L, Liu J (2013) Adaptive boundary control for flexible two-link manipulator based on partial differential equation dynamic model. Control Theory Appl IET 7(1):43–51

Chapter 10
Conclusion

The book has been dedicated to the modeling and control design of flexible mechanica systems. The book is divided into 10 chapters, and modeling of three typical flexible mechanical system including a flexible satellite based on fourth order PDE, a flexible aerial refueling hose based on second order PDE in two dimensions, and a flexible manipulator based on fourth order PDE in three dimensions, and a variety of control methods are introduced. The results of the research work conducted in this book are summarized in each chapter, and the contributions made are reviewed. The key results are listed as follows.

In Chap. 3, the modeling and control problem of a satellite with flexible solar panels has been addressed by using a single-point control input. The panels with flexibility have been modeled as a distributed parameter system described by hybrid PDEs-ODEs. The control input has been proposed on the original PDE dynamics to suppress the vibrations of two panels. Then exponential stability has been proved by introducing a proper Lyapunov function. The effectiveness of the proposed control has been verified by simulations.

In Chap. 4, we study vibration control to stabilize the flexible satellite described by a distributed parameter system modeled as PDEs with input constraint and external disturbance. Backstepping method is used to regulate the vibration of the flexible satellite in the presence of external disturbance and constrained input. In the controller design, an auxiliary system based on a smooth hyperbolic function and a Nussbaum function is adopted to handle the impact of the external disturbance and constrained input. The Lyapunov function is applied for control law design and stability analysis of the close-loop system. Numerical simulations have been provided to illustrate the effectiveness of the proposed boundary control.

In Chap. 5, we addresses the modeling and boundary control problem of a flexible aerial refueling hose. The flexible hose system is modeled as a distributed parameter system. Then a boundary control scheme is proposed based on Lyapunov's direct method to regulate the hose's vibration. A disturbance observer is designed to estimate the input disturbance. With the proposed control and the disturbance observer, the close-loop stability is proven through rigorous analysis without any simplification or discretization of the partial differential equation (PDE) dynamics.

© Tsinghua University Press 2020
Z. Liu and J. Liu, *PDE Modeling and Boundary Control for Flexible
Mechanical System*, Springer Tracts in Mechanical Engineering,
https://doi.org/10.1007/978-981-15-2596-4_10

In Chap. 6, the boundary control problem of a flexible aerial refueling hose modeled as a distributed parameter system with input saturation is addressed. A boundary control scheme is proposed based on backstepping method to regulate the hose's vibration. An auxiliary system based on a smooth hyperbolic function and a Nussbaum function is designed to handle the effect of the input saturation. With the Lyapunov's direct method, the closed-loop stability is proven through rigorous analysis without any simplification or discretization of the partial differential equation (PDE) dynamics. The advantages of the proposed control scheme are that: (1) it can deal with the vibration of the hose with input saturation, the external disturbance and the horizontal velocity; and (2) the close-loop stability analysis avoids any simplification or discretization of the PDEs based on the Lyapunov's method.

In Chap. 7, we design a boundary control to stabilize a flexible hose modeled as a distributed parameter system (DPS) with input deadzone and output constraint. In the controller design, a radial basis function (RBF) neural network is used to handle the effect of the input deadzone, and a barrier Lyapunov function is employed to prevent constraint violation. The Lyapunov approach is applied for the stability analysis of the close-loop system. The numerical simulations verify the effectiveness of the presented method.

In Chap. 8, we addresses the modeling and control problem of a flexible aerial refueling hose with varying length, varying speed, and input constraint. A boundary control scheme is proposed based on backstepping method to regulate the hose's vibration. To the best of our knowledge, this is the first application of the boundary control design for a flexible hose system with varying length and input constraint based on original PDEs. In the controller design, an auxiliary system based on a smooth hyperbolic function and a Nussbaum function is designed to handle the effect of the input constraint. With the Lyapunov's direct method, the closed-loop stability is proven through rigorous analysis without any simplification or discretization of the partial differential equation dynamics. In the simulation study, by appropriate choices of the control parameters, the vibration of the flexible hose is suppressed and the effect of input constraint is handled.

In Chap. 9, the dynamic equations which indicates strong nonlinear coupling for a three-dimensional flexible manipulator are derived, then used for the design of the boundary control. The disturbance observers are proposed to estimate the input disturbances. Based on the observers, boundary control schemes are designed based on the Lyapunov's direct method and the original PDEs and ODEs model. With the proposed control, orientation can be regulated and deflections suppressed simultaneously. The advantages of the proposed control scheme are that: (i) it can deal with the vibration of the flexible manipulator with input disturbances; and (ii) the close-loop stability analysis avoids any simplification or discretization of the PDEs based on the Lyapunov's method.

In summary, this book covers the dynamical analysis and control design for three typical flexible mechanical systems, which can be applied to other mechanical systems. The book is primarily intended for researchers and engineers in the control system. It can also serve as a complementary reading on modeling and control of flexible mechanical systems at the post-graduate level.

Printed in the United States
by Baker & Taylor Publisher Services